Mircea Negoita, Daniel Neagu, Vasile Palade

Computational Intelligence: Engineering of Hybrid Systems

Studies in Fuzziness and Soft Computing, Volume 174

Editor-in-chief
Prof. Janusz Kacprzyk
Systems Research Institute
Polish Academy of Sciences
ul. Newelska 6
01-447 Warsaw
Poland
E-mail: kacprzyk@ibspan.waw.pl

Mircea Negoita
Daniel Neagu
Vasile Palade

Computational Intelligence: Engineering of Hybrid Systems

 Springer

Professor Mircea Gh. Negoita
Wellington Institute of Technology
School Information Technology
Centre for Computational Intelligence
Te Puni Mail Centre Buick Street
Private Bag 39803, Wellington
New Zealand
E-mail: mircea.negoita@weltec.ac.nz

Dr. Vasile Palade
University of Oxford
Computing Lab
Parks Rd, Wolfson Building
Oxford, OX1 3QD
U.K.
E-mail: Vasile.Palade@comlab.ox.ac.uk

Dr. Daniel Neagu
University of Bradford
Department of Computing
Bradford, West Yorkshire, BD7 1DP
U.K.
E-mail: D.Neagu@bradford.ac.uk

ISBN 978-3-642-06224-7   e-ISBN 978-3-540-32369-3

ISSN print edition:1434-9922
ISSN electronic edition: 1860-0808

This work is subject to copyright. All rights are reserved, whether the whole or part of the material is concerned, specifically the rights of translation, reprinting, reuse of illustrations, recitation, broadcasting, reproduction on microfilm or in any other way, and storage in data banks. Duplication of this publication or parts thereof is permitted only under the provisions of the German Copyright Law of September 9, 1965, in its current version, and permission for use must always be obtained from Springer. Violations are liable for prosecution under the German Copyright Law.

Springer is a part of Springer Science+Business Media
springeronline.com

© Springer-Verlag Berlin Heidelberg 2005
Softcover reprint of the hardcover 1st edition 2005

The use of general descriptive names, registered names, trademarks, etc. in this publication does not imply, even in the absence of a specific statement, that such names are exempt from the relevant protective laws and regulations and therefore free for general use.

Cover design: E. Kirchner, Springer Heidelberg
Printed on acid-free paper   62/3141/jl- 5 4 3 2 1 0

To erudite professor and scholar Severin Bumbaru

To our lovely and patient wives Doina Negoita,
Daniela Neagu and Dana Palade

Invited Preface

It is my great pleasure to introduce to you this excellent book on computational intelligence written by three well-known researchers: Mircea Gh. Negoita of New Zealand, and Ciprian Daniel Neagu and Vasile Palade of Great Britain. The authors have published extensively in the area of computational intelligence and the book benefits greatly from their experience.

Computational intelligence is a very broad term meaning different things to different people. However, recent creation of the Computational Intelligence Society, by the Institute of Electric and Electronic Engineers (IEEE), and the scope of journals it publishes, enables us to roughly define computational intelligence as one encompassing neural networks, fuzzy systems, evolutionary computation, their variants and hybrids.

It is the hybrid approaches, combining several methodologies that are paid special attention in the book. Many indicators show that such hybrid intelligent systems are one of the fastest growing fields that attract hundreds of researchers and practitioners. It comes as no surprise that hybrid approaches are required on dealing with real-world problems since (artificial) intelligence cannot be accomplished by any single methodology alone. There are many views on human intelligence as well, and it can also be seen as a hybrid of several types of "intelligence".

It is the view of the authors that hybrid intelligent systems are more than just the three broad areas mentioned above. Consequently, they also cover knowledge extraction and fusion, and novel areas such as DNA computing, Artificial Immune Systems and Evolvable Hardware.

The book presents in a thorough way almost an entire range of hybrid intelligent systems, and what is very important for practitioners, it also provides several real-life, working, implementations of hybrid systems.

The book probably is the first attempt to provide the basis for hybrid intelligent systems through clear and well-organized way of presenting the fundamentals of key methods. As a result the book can serve as a valuable introduction to hybrid intelligent systems, and as a guide of how to use them for solving real problems. As such it has a good chance of becoming one of the most cited references in the area.

I am convinced that the book will play an important role in pursuing further developments and applications of hybrid intelligent systems methods.

The authors well deserve congratulations for their excellent work, and I commend to the readers the volume that is an important addition to the literature of computational intelligence.

University of Colorado at Denver   *Krzysztof Cios*
and Health Sciences Center

Preface

The *Soft Computing* framework or *Computational Intelligence* with its large variety of efficient applications is hugely fascinating. Problems in engineering, computational science and the physical and biological sciences are using the increasingly sophisticated methods of *Computational Intelligence*. Because of the high interdisciplinary requirements featuring most real-world applications, no bridge exists between the different stand alone *Intelligent Technologies*, namely Fuzzy Systems, Neural Networks, Evolutionary Algorithms, DNA Computing, Artificial Immune Systems and Knowledge Based Systems. The concomitant increase in dialogue and interconnection between the *Intelligent Technologies* has led to *Computational Intelligence* and its practical implementation – the *Hybrid Intelligent Systems*. The idea of writing a book on this topic first crossed my mind in 1996, and I am really happy that the book is finally complete.

Initially I thought this book would be of real help to my gifted students at the Department of Applied Electronics and at the Department of Automatics in the Faculty of Electrical Engineering and Navy of the University "Dunarea de Jos" Galati, Romania. New ideas and suggestions regarding the final structure of the book were obtained from students attending my course on Intelligent Multi-Agent Hybrid Systems at Wellington Institute of Technology, New Zealand. They were really pleased to become familiar with *the Intelligent Technologies* and with some typical competitive software, such as Neural Network Professional II Plus and Soft Computing Genetic Tool. Their interest in this area gave me the stimulus to finish the book. My intention was to be helpful to the students, not to exempt them from intellectual effort, but to put as much illustrative information as possible into the book. The purpose was to create *a very clear image* of what *Computational Intelligence* is in general terms, and to convince them that *Hybrid Intelligent Systems* are really nothing else but the engineering implementation of *Computational Intelligence*.

The purpose of this book is to illustrate the current needs and to emphasize the future needs for the interaction between various *Intelligent Technologies*. The team writing this book did this firstly by encouraging the ways that *Intelligent Technologies* may be applied in those areas where they are already traditional, as well as pointing towards new and innovative areas of application involving emergent technologies such as DNA Computing, Artificial Immune Systems and Evolvable Hardware. Secondly, to help encourage other disciplines to engage in a dialogue with practitioners of *Hybrid Intelligent Systems* engineering, outlining their problems in accessing these new methods in the engineering of *Hybrid Intelligent Systems*, and also suggesting innovative developments within the area itself.

An important concept in my university teaching/research work was that a student must be convinced at an early professional stage that there is an interdisciplinary aspect in the development of technology. Students require not only a knowledge/skills base, but also the methodology for implementing real world applications under interdisciplinary conditions. Nowadays this is available not only to recent graduates, but also to other practitioners (engineers, scientific researchers and academic staff) whatever their background and area of interest may be. Thus the progress of *Hybrid Intelligent Systems* within the framework of *Computational Intelligence* was discussed from an application – engineering point of view, rather than from a cognitive science or philosophic view point. In our book we have not detailed all aspects of the *Computational Intelligence* framework as viewed by its founder L. A. Zadeh. The second feature of this book is a discussion of the interactive aspects of different *Intelligent Technologies* which have lead to both the evolution and practical interactions of *Hybrid Intelligent Systems*.

Special attention (separate book chapters) is given to the interaction of the I*ntelligent Technologies* with DNA Computing, Artificial Immune Systems and with the most spectacular emergent technology – **E**volvable **H**ard**W**are *(EHW)*. *EHW* has opened a revolutionary era in technology and in the social life development of humans by its radical impact on engineering design and automation. A dream of humanity has become reality: *EHW* has transferred the adaptivity of a system from software to hardware. A significant time-saving path is used from a design to a real world application of intelligent hardware. Differences no longer exist between the design and adaptation involving *EHW*-based machines having behavioural computational intelligence. Electronic engineering has been fundamentally changed by using *EHW* custom-design technologies instead of solder based manufacturing.

Chapter 1 is an introduction to *Computational Intelligence* and *Hybrid Intelligent Systems*, including their terminology and classification.

Chapter 2 presents a major application of *Hybrid Intelligent Systems* – fault diagnosis, and is focused mainly on neural-fuzzy methods.

Chapter 3 describes theoretical aspects of the main neural-fuzzy techniques for the design of *Hybrid Intelligent Systems*, but hybrid intelligent systems combining connectionist and symbolic features are also presented.

Chapter 4 focuses on the strictly fuzzy approach of neural networks, and presents some interactive fuzzy operators in order to extract connectionist-represented knowledge using the concept of f-duality. The methodologies are tested on simple and traditional case studies.

Chapter 5 is an introduction to modular networks in fuzzy systems. This provides new insights into the integration of explicit and implicit knowledge in a connectionist representation.

Chapter 6 is focussed on application aspects of *Hybrid Intelligent Systems* engineering. The main application areas of *Hybrid Intelligent Systems* are mentioned. Some special original applications are introduced. *NEIKeS* (**N**eural **E**xplicit and **I**mplicit **K**nowledge Based **S**ystem) is an original Fuzzy Knowledge-Based System for the prediction of air quality. *WITNeSS* (**W**ellington **I**nstitute of **T**echnology **N**ovel **E**xpert **S**tudent **S**upport) is an intelligent tutoring system. Two applications of *VLGGA* (**V**ariable **L**ength **G**enotype **G**enetic **A**lgorithm) – a special non-standard *GA* – are introduced in the form of two hybrid intelligent optimisation methods which have wide applicability.

The most recent trends in the development of *Hybrid Intelligent Systems* are presented in *Chap. 7*. These applications rely on the hybridization techniques of DNA Computing and Artificial Immune Systems.

Chapter 8 emphasizes the massive role played by Evolutionary Computation in the implementation of *Computational Intelligence*. A large range of GA based *Hybrid Intelligent Systems* are introduced, with applications in Fuzzy information processing, and Neural Networks design and optimization. The role of *Hybrid Intelligent Systems* in the design of high performance GA is illustrated. A special part of this chapter is reserved for GA based *Hybrid Intelligent Systems* in EHW implementation.

Most sections include useful suggestions for the practical design and development of further applications. All three authors agree that although this book is a primer, it is not useful to only students. This book has practical value for both those new to the discipline and also for those who are already practitioners in the area.

The common research work with my research team colleagues while I was at the University of Galati, Romania – Dr. Ciprian Daniel Neagu (Bradford University, UK) and Prof. Vasile Padae (Oxford University, UK) – constitutes the foundation for this book. Some of the results in their PhD theses are introduced in this book. The time which I spent abroad both as a visiting professor and as a visiting researcher at Dortmund University (Germany), at York University (Toronto, Canada), at RWTH (Aachen, Germany), and the permanent connection and exchange of ideas with colleagues from abroad, were of great value in obtaining the original results that are presented here. Some common research developments with my brilliant former research team-mates (Alexandru Agapie, Florin Fagarasan, Marius Giuclea and Dan Mihaila) from National Institute for Research and Development in Microtechnologies (IMT) in Bucharest, and also some very recent results of international scientific cooperation with Dr. Dragos Arotaritei from Aalborg Univesity (Esbjerg, Denmark) are integrated into some sub-chapters.

I have just one regret regarding this book: Prof. Edmond Nicolau, world renowned as one of the co-founders of modern cybernetics, former Director in World Organization of Cybernetics and Systems, a pioneer of scientific research in Neural Networks, my PhD supervisor and scientific mentor passed away before this book was finished.

A decisive element for finally completing the book was the environment and research conditions I met at Wellington Institute of Technology (WelTec), Wellington, New Zealand. Special thanks are due to Dr. Linda Sissons, WelTec CEO and to Denford McDonald, WelTec Council Chairman, for understanding the important role of *Computational Intelligence* in economic development and for giving me their full support in my work as the Director of Centre for Computational Intelligence at WelTec.

The emotional feelings involving the completion of this book which I have written in New Zealand – together with my best research colleagues – were that I never forget my native country – *Romania*, but I deeply love my adoptive country – *New Zealand*.

The whole team of authors is grateful for the understanding and permanent support of Springer Verlag Publishing House throughout the writing of this book. We would also like to acknowledge our special appreciation for the hard work done by David Pritchard, my research team colleague at WelTec Centre for Computational Intelligence, and, being a native English language speaker, for finally "brushing" the camera-ready manuscript. Special acknowledgments must be addressed to Professor Nikolaos Avouris and the Human Computing Interaction Group of the University of Patras, Greece, where the prototype for the air quality prediction system NEIKeS has been developed. Also we acknowledge the support of the EU FP5 RTN project IMAGETOX, Professor Giuseppina Gini, Dr. Emilio Benfenati and Dr. Mark Cronin for the collaborative research training network the prortotype of the system NIKE for Predictive Toxicology has been developed.

Wellington, New Zealand   *Mircea Gh. Negoita*
January 2005   *Daniel Neagu*
                                                                Vasile Palade

Contents

List of Acronymes

A	*Adenine (a **DNA** base)*
AI	*Artificial Intelligence*
aiNet	*Artificial Immune Network*
AIS	*Artificial Immune Systems*
AL	***Aggregative** Layer*
AN	*Adaptive Networks*
ANN	*Artificial Neural Networks*
ANFIS	*Adaptive Neuro-Fuzzy Inference Systems*
BCA	*B-cell Algorithm*
BPNN	*Back Propagation Neural Networks*
C	*Cytosine*
CAI	*Conexionist Artificial Intelligence*
CI	*Computational Intelligence*
CRI	*Compositional Rule of Inference*
CRF	*Combine the Rules First*
CSP	*Connectionist Symbol Processing*
CMSE-OLC	*Constrained Mean-Squared Error-Optimal Linear Combination Algorithm*
CSM	*Concept Support Method*
CWP	*Computing with Words and Perceptions*
DCPS	*Distributed Connectionist Production System*
DDS	*Distributed Database System*
DFRBS	*Discrete Fuzzy Rule Based System*
DNA	*DeoxyriboNucleic Acid*
DNA-AIS	*Hybridization between DeoxyriboNucleic Acid systems and Artificial Immune Systems*
DNA-FS	*Hybridization of DeoxyriboNucleic Acid techniques and Fuzzy Systems*
DNA-NN	*Hybridization of DeoxyriboNucleic Acid techniques and Neural Network techniques*
DNA-GA	*Hybridization of DeoxyriboNucleic Acid techniques and Genetic Algorithms*
DPGA	*Dynamic Parameter Genetic Algorithm*
EA	*Evolutionary Algorithms*

EC	*Evolutionary Computation*
EC-AIS	*Hybridization between Evolutionary Computation techniques and Artificial Immune Systems*
EHW	*Evolvable HardWare*
EHW-AIS	*Hybridization between Evolvable HardWare and Artificial Immune Systems*
EK	*Expert Knowledge*
EKM	*Explicit Knowledge Module*
EMSE-OLC	*Estimated Mean-Squared Error-Optimal Linear Combination Algorithm*
EP	*Evolutionary Programming*
ES	*Expert Systems*
FAS	*Fuzzy Additive System – as described by TSK rules*
FGA	*Fuzzy Genetic Algorithms*
FPAA	*Field Programmable Analogue Arrays*
FPGA	*Field Programmable Gate Arrays*
FS	*Fuzzy Systems*
FS-AIS	*Hybridization between Fuzzy Systems and Artificial Immune Systems*
FCM	*Fuzzy-c-Means algorithm*
FDI	*Fault Detection and Isolation*
FEM	*Fire Each Module*
FIS	*Fuzzy Inference System*
FNN	*Fuzzy Neural Network*
FRBS	*Fuzzy Rule-Based Systems*
FR	*Fuzzy Rule*
FRF	*Fire the Rules First*
FSM	*Finite State Machine*
G	*Guanine*
GA	*Genetic Algorithms*
GA-FS	*Genetic Algorithms hybridization with Fuzzy Systems*
GA NN	*Genetic Algorithms hybridization with Neural Networks*
GA-NN-FS	*Genetic Algorithms-Neural Networks-Fuzzy Systems*
HIS	*relied Hybrid Intelligent System*
GMP	*Generalized Modus Ponens*
GP	*Genetic Programming*
GSO	*Global Selection Operator*
HFNN	*Hybrid Fuzzy Neural Network*
HIS	*Hybrid Intelligent System*
HME	*Hierarchical Mixtures of Experts*
HNN	*Hybrid Neural Network*
HNS	*Hybrid Neural System*
HTN	*High Treshold Neuron*

IKM	*Implicit* **K**nowledge **M**odule
HIS	**H**ybrid **I**ntelligent **S**ystem
IP	**I**nternet **P**rotocol
IT	**I**ntelligent **T**echnologies
ITS	**I**ntelligent **T**utoring **S**ystem
KBANN	**K**nowledge-**B**ased **A**rtificial **N**eural **N**etworks
KBES	**K**nowledge-**B**ased **E**xpert **S**ysyem
KBS	**K**nowledge-**B**ased **S**ystem
KDD	**K**nowledge **D**iscovery *in* **D**atabases
K-MCA	**K**-**M**eans **C**lustering **A**lgorithm
LMS	**L**east **M**ean **S**quares *(Widrow) algorithm*
LOP	**L**ogarithmic **O**pinion **P**ool
LSE	**L**east-**S**quares **E**stimator
LTN	**L**ow **T**reshold **N**euron
MAPI	**M**atching, **A**ggregation, **P**rojection, **I**nverse **F**uzzification
MAX	*fuzzy* **max***im opearor*
MC	**M**olecular **C**omputing
MF	**M**embership **F**unction
MIMO	**M**ulti-**I**nput **M**ulti-**O**utput
MIN	*fuzzy* **min***im opearor*
MILA	**M**ultilevel **I**mmune **A**lgorithm
MISO	**M**ulti-**I**nput **S**ingle-**O**utput
MLP	**M**ulti-**L**ayer **P**erceptron
MPNN	**M**ulti**P**urpose **N**eural **N**etwork
MSE	**M**ean-**S**quared **E**rror
MSE-OLC	**M**ean-**S**quared **E**rror-**O**ptimal **L**inear **C**ombination *algorithm*
MultiNN **Prom**	**M**ulti**p**le **N**eural **N**etwork *Based System for* **Prom***oter Recognition*
N	**N**eural *(neural component of an HIS)*
NN	**N**eural **N**etwork
NES	**N**eural **E**xpert **S**ystems
NEIKeS	**N**eural **E**xplicit *and* **I**mplicit **K**nowledge-*based* **S**ystem
NGA	**G**enetic **A**lgorithm *of* **N**agoya *type*
NKB	**N**eural **K**nowledge **B**ase
NNR	**N**earest-**N**eighbour **R**ule
NN-FS HIS	**N**eural **N**etwork – **F**uzzy **S**ystems *based* **H**ybrid **I**ntelligent **S**ystem
NRFC	**N**ew **F**uzzy **R**easoning **F**uzzy **C**ontrollers
NSP	**N**euronal **S**ymbol **P**rocessing
OLC	**O**ptimal **L**inear **C**ombination
OLS	**O**rthogonal **L**east-**S**quares *algorithm*
PI	**P**roportional **I**ntegral
PPX	**P**recedence **P**reserving **C**rossover

PPS	*Precedence Preserving Mutation*
PRODUCT	*fuzzy **product** operator*
QSAR	*Quantitative Structure-Activity Relationship*
RAFNN	Recurrent Artificial Neural Network with Fuzzy Numbers
RAINN	*Resource limited Artificial Immune Network*
RBFN	Radial Basis Function Networks
RBS	Rule-Based Systems
RHIS	Robust (Soft Computing) Hybrid Intelligent Systems
RLS	Recursive Least-Squares algorithm
RLSE	Recursive Least-Squares Error
RNN	Rule Neural Network
RRS	Relative Rule Strength
S	*Symbolic (symbolic component of an HIS)*
SAI	Symbolic Artificial Intelligence
SAR	*Structure-Activity Relationship*
SC	Soft Computuing
SGA	Simple Genetic Algorithm
SGN	Supervised-trained Gating Network
SO	Student Object
SOM	Self Organizing Maps *(Kohonen Neural Networks)*
T	Thymine
TNGS	*Theory of Neuronal Group Selection*
TP	Triangular Partition
TPE	Triangular Equidistant Partition
TPh	Triangular Partition with Variable Length (hyperbolic sements)
UGN	Unsupervised-trained Gating Network
U MSE-OLC	Unconstrained Mean-Squared Error-Optimal Linear Combination algorithm
VLGGA	Variable Length Genotype Genetic Algorithm
WITNeSS	Wellington Institute of Technology Novel Expert Student Support
XOR	*EXclusive **or** operator*

1 Intelligent Techniques
and Computational Intelligence

The field of **I**ntelligent **T**echnologies (*IT*) or **C**omputational **I**ntelligence (*CI*) is mainly the result of an increasing merger of **F**uzzy **S**ystems (*FS*) or **F**uzzy **L**ogic (*FL*), **N**eural **N**etwork (*NN*) and **E**volutionary **C**omputation (*EC*). These technologies are providing *increasing benefit to business and industry*. Most of the countries in the world have understood the important role of *CI*. Not only have governmental institutions support National Projects in this area, but a lot of private companies are also currently using intelligent technologies in a number of application areas. For example *Siemens*, *Fisher & Paykel* use *FS* in different *intelligent controllers* for household appliances and New Zealand Paraparaumu *Sewage Plant* uses *NN* for inflow *prediction*). These applications are not the consequence of fashionable, trendy ideas but a response to real world needs (Negoita 2002; Negoita 2003).

These technologies have the power to *increasing improveg any area of our social-economic life*. Nowadays we are confronted with a lot of complex real-world applications. Examples would be different forms of pattern recognition (image, speech or handwriting), robotics, forecast and different kind of decision-making in uncertainty conditions, etc... These kind of applications reply on a new concept, a new framework that is named **S**oft **C**omputing (*SC*) by L.A. Zadeh, the father of fuzzy systems. The basic ideas underlying *SC* belong to Prof. Zadeh. His vision of different techniques of interaction is strongly influencing the development of the new generation of intelligent (perception-based) systems toward a **C**omputational **I**ntelligence (**CI**).

New tools are developing for dealing with world knowledge, for example the *FL*-based method of **C**omputing with **W**ords and **P**erceptions (*CWP*) featured by the understanding that perceptions are described in a natural language (Zadeh and Nikravesh 2002; Zadeh 2003).

1.1 "Computational" and "Artificial" Intelligence Distinguished

A large bunch of methods have been developed as useful tools of information processing, machine learning and knowledge representation. These methods are either an outcome of the "**A**rtificial *Intelligence*" (*AI*) or "**I**ntelligent **T**echnologies" (*IT*) frameworks. The *AI* framework relies mainly on symbolic

Mircea Gh. Negoita, Daniel Neagu, and Vasile Palade: *Computational Intelligence: Engineering of Hybrid Systems*, StudFuzz **174**, 1–11 (2005)
www.springerlink.com   © Springer-Verlag Berlin Heidelberg 2005

Knowledge **B**ased **E**xpert **S**ystems (*KBES*) while the *IT* framework relies mainly on *FS, NN, EC* and even on some other unconventional methods as *DNA* computing. The *DNA* computing framework was enlarged by *DNA* hybridisation. *FS* was implemented in *DNA* bio molecules, *DNA NNs* became available and *DNA* computing began to be applied in the implementation of **G**enetic **P**rogramming (*GP*).

But *IT* and *AI* are distinct frameworks and must be distinguished with the aim of improving their exploitation in applications. Let us make a comparison between the *IT* and *AI* frameworks, focussing on four main characteristics: the main purpose, the methodological framework, the nature of the information processing and the proven, successful applications typical to each of them (see Table 1.1 for the details regarding this comparison).

Table 1.1. A Comparison between AI and CI

	AI	IT (CI)
Main purpose	Conceiving and design of *intelligent non-living systems*	Increasing the machine intelligence quotient of non-living systems (giving the machines the ability to *think* and manifest behaviour that is as close as possible to *the behaviour of human beings*)
Methods	Classical ones, most based on *first-order bivalent predicate logic*	Solutions of the problems are satisfactory in form of *global approximations* but in most cases *they are not optimal*
Information processing	By handling *symbols*	By *numerical computation*
Some typical successful applications	• Knowledge-based engineering • Fault diagnosis, etc...	• Different forms of pattern recognition (image, speech, handwriting) • Robotics • Forecast applications • Decision-making under uncertain conditions, etc...

Each *IT* and *AI* method has its particular *strengths* and *weaknesses* that make them suitable for some applications and not other. These specific limitations of *AI* and *IT* led to creation of **H**ybrid **I**ntelligent **S**ystems (*HIS*), where two or *more techniques are combined* and co-operate to overcome the limitations of each individual technique. An overview of the main individual *IT* and *AI* techniques will be useful with regard to their successful applications, the advantages and limitations manifesting the individual effectiveness of the method.

FS deals with the handling of imprecision in data, finally making a decision made by *fuzzy concepts* and *human like approximate reasoning*. *FS* have knowledge bases that, because they are based on fuzzy IF-THEN rules, are easy to examine, understand, update and maintain. Another typical advantage of *FS* is their high degree of flexibility in dealing with incomplete and inconsistent data by aggregating the hypothesis of all the rules. The scope of successful *FS* applications is large and diverse. To name just a few, in industry with different intelligent microcontrollers; transportation with the control of rail, highway and air traffic and in financial banking with exchange prediction or stock market analysis. But *FS* also has real limitations – the heavy *reliance on human experts* in building the fuzzy IF-THEN rules, the need for *manual specification* of membership functions and fuzzy rules, as well as strong restrictions in *automatic adaptive learning* capabilities in an ever changing external environment.

NN deals with inherent *parallelism* in data flow. Their capacity to learn patterns in data that are noisy, incomplete and even contradictory, confer on *NN* two distinct advantages in processing information – the ability of *processing incomplete information* and *generalize the conclusions*. The application of *NN* is even larger in scope than *FS*. In the financial/business sector we have the prediction and modeling of markets, assessments of credit worthiness and selection of investments, the automatic reading of handwritten characters (cheques) and signature analysis. In telecommunications we have signal analysis, noise elimination, and data compression, while in the environmental sector there is risk evaluation, chemical analysis, weather forecasting, resource management. In industry we have quality control and production planning and in medicine the medical diagnosis. These are just a few of the applications of *NN*. *NN* effectiveness is due to the easy modeling and forecasting of non-linear systems and its ability to handle tasks involving incomplete data, unavailable expert advice or where no rules can be formulated. The same as *FS* has its limitations, so does *NN*. *NNs* are *unable to explain how they arrived at their conclusions* and are unable to interact with conventional data bases. Also there is a *lack of structured knowledge representation*. Also *NN* have real *scalability problems* in that they experience great difficult in training and in generalization capabilities for large and complex problems.

Like *NN, EC* (**G**enetic **A**lgorithms – *GA*, especially) also deal with *parallelism* in data flow. *EC* provides efficient search and optimization tools for very large data sets, largely unconstrained by local minima problems. They don't require any mathematical modeling. *EC* are optimization methods for either symbolic or *NN* or *FS* systems. Strictly referring to *GA*, their effectiveness consists of a high suitability for parallel computer implementation;

particular success in large search and optimization problems; in an ability to learn complex relationships in incomplete data sets, their use as "data mining" tools for discovering previously unknown pattern; in their ability to provide an explanation of how decisions were produced, in a format that humans can understand; an ability to adapt to changes in their operating environment. *GA* limitations are as follows: they are *computational expensive* methods involving an *application dependent setting* of *GA* parameters (e.g. for crossover, mutation) that could be a time-consuming trial and error process. Typical successfully application to *GA* are many. The generation of control strategies for industrial processes (water treatment systems, cascaded chemical reactors, crackers, fermenters, distillatory columns, etc...). There is the optimization of economic dispatch problem and unit commitment problem in electric power production. The use of optimized truck routing in delivery systems. Optimization of manufacturing processes and job shop scheduling in industry. Aerospace applications use *GA* in the attitude determination of a spacecraft. There is the design of antenna and electronic circuits, including transducers. *GA* is also used in developing tailor-made solutions to individual customer requirements. And last, but not least, real-time adaptive hardware (**E***volvable* **H***ardware* – *EHW*).

DNA computing is certainly a promising implementation method of information-processing capabilities of organic molecules, used with the aim of replacing digital switching primitives in computers (Paun et al. 1998). This kind of computing relies strongly on the availability of a proper working environment (technological support devices). Although a jump from microchip electronic devices to *DNA* molecules is the final expectation, in reality the technology is still struggling to combine electronic computing with *DNA* computing, where most of the *DNA* operation are still being carried out in test tubes without the intervention of the user. *DNA* computing is characterized by a massive parallelism conferred by *DNA* strands which means the first strength of *DNA* computing is much *higher parallel data structures* than either *NN* or *GA*. The second strength of *DNA* computing devices results from the *complementarity* between two *DNA* strands when bonding takes place. Already proved successfully application include cryptanalysis of a cipher text and some robotics elements. DNA effectiveness as a powerful tool for computing relies on the handling of suitable encoded information of a very high density that leads to far-reaching conclusions when bonding takes place. The main limitation of *DNA* computing consists of a *higher than admissible error rate of operations* due to an actual lack of a proper technological environment.

1.2 Key Features of Intelligent Systems

It was the computational and practical issues of real world-applications, highlighting the strengths and weaknesses of *CI* and *AI* methods that led to evolution and development of *HIS*. The evolution of *HIS* is a consequence of modelling human information processing, but it was the interdisciplinary vision of solving a large range of real-world problems that resulted in application engineering hybridizing the CI and AI methods. A key factor of the success in effectively using *CI* and *AI* methods is a real understanding of the strengths and applicability of *HIS* in the very large industrial and finance/business environment. Much more, the specialized personnel in *HIS* software engineering acts as a complex and complete social development agent, a so-called "business engineer" augmenting the *CI* information processing by the addition of typical knowledge elements to *AI*. So non-living systems are modelled on the biology of human intelligence by the integration of both *CI* and *AI* methods.

In conclusion, the development of *HIS* application is justified in that no *CI* or *AI* method can be applied universally to every type of problem, each technique having its advantages and disadvantages. There are some *key features* that most *HIS* display, which make them particularly useful for problem solving and certainly no *CI/AI* technique exhibits *all these features*. The *key (application required) features* making *HIS* development suitable for use with applications are *learning, adaptation, flexibility, explanation* and *discovery* (Goonatilake and Treleaven 1996). Accurate quantitative methods would be ideal for assessing the potential availability of the main *CI/AI* techniques in providing the five main desired *HIS* properties. No such tools are available, but a qualitative balance using the remarks accumulated and proved as a result of application development might perform the analysis.

An *HIS* must be able of performing a simple mining through a huge amount of previously acquired records of input data, to arrive at a model of the application. This *HIS* feature is called *learning* and means the capability of learning the tasks/decision that *HIS* has to perform, directly from the collected data. See Fig. 1.1 for a qualitative balance of the main *CI/AI* techniques with respect to their *learning* abilities.

Fig. 1.1. A qualitative balance of the main *CI/AI* techniques with respect to their *learning* abilities

An *HIS* must be able to express the intelligence of having decision-making procedures that execute in way that can be understood by humans. These decision-making procedures must be transparent, allowing the reasoning process to be understood and modified to improve the *HIS*. This *HIS* feature is known under the name of *explanation*. See Fig. 1.2 for a

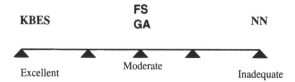

Fig. 1.2. A qualitative balance of the main *CI/AI* techniques with respect to their *explanation* abilities

qualitative balance of the main CI/AI techniques with respect to their *explanation* abilities.

But learning is a process that must continuously feature the activity of any *HIS* in any conditions (learning just an initial amount of knowledge is not at all enough for making an *HIS* able of performing a particularly task). *HIS* must be able to monitor the system tasks constantly (including knowledge revision according to any change in the operating environment). This HIS feature is called *adaptation*. See Fig. 1.3 for a qualitative balance of the main CI/AI techniques with respect to their *adaptation* abilities.

Fig. 1.3. A qualitative balance of the main *CI/AI* techniques with respect to their *adaptation* abilities

The *HIS* must be able to perform decision making even when there is imprecise, incomplete or completely new input data. This basic *HIS* feature is called *flexibility*. See Fig. 1.4 for a qualitative balance of the main *CI/AI* techniques with respect to their *flexibility* abilities.

Discovery is *HIS* data mining ability – the capability of mining through a huge amount of collected input data and not only finding relationships that were previously unknown, but checking to see whether the discoveries were

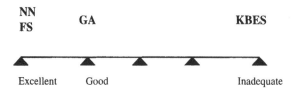

Fig. 1.4. A qualitative balance of the main *CI/AI* techniques with respect to their *flexibility* abilities

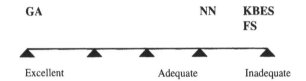

Fig. 1.5. A qualitative balance of the main *CI/AI* techniques with respect to their *discovery* abilities

not just statistical flukes. See Fig. 1.5 for a qualitative balance of the main *CI/AI* techniques with respect to their *discovery* abilities.

No accurate quantitative comparison methods are available for assessing the potential availability of the main *CI/AI* techniques providing the five main desired *HIS* properties. But a comparison might be performed using the experience accumulated and proved as a result of application development (see Table 1.2).

Table 1.2. Potential availability of main *CI/AI* techniques in providing the five main desired *HIS* properties

CI/AI Technique	Learning	Explanation	Adaptation	Discovery	Flexibility
KBS	inadequate	excellent	inadequate	inadequate	inadequate
FS	inadequate	moderate	inadequate	inadequate	excellent
NN	excellent	inadequate	excellent	adequate	excellent
GA	excellent	moderate	good	excellent	good

Some comments must be made on the compatibility of main *CI/AI* methods to an ideal implementation, namely for conferring to *HIS* the typical features of real-time systems. Real-time systems are systems that function under very sharp (application required) time and space constraints. For example some of the usual time constraints include response/transition times and reasoning under time constraints. There are still a lot of practical issues that drastically limit the availability of the main *CI/AI* methods to provide *HIS* with the required real-time qualities (including high reliability). It is clear that greater technological support is needed. It is a realistic claim that the main *CI/AI* methods presently available are still not good enough to meet some applications requirements for special time, space and even sharp logical constraints. Severe consequences might result if logical as well as timing correctness properties of a real time system are not satisfied.

1.3 Classification of Hybrid Intelligent Systems

Section 1.2 has mentioned the *key (application required) features* making *HIS* development suitable for one or other application, as the case. The summarized quantitative comparison, as in Table 1.2, is to be used by practitioners as the starting point in building *HIS* that are effective in applications. Table 1.2 suggests two practical strategies in building *HIS* and they are classified according to these design strategies (Goonatilake and Treleaven 1996). This classification divides *HIS* into two classes: *function–replacing HIS* and *intercommunicating HIS*.

Function–replacing HIS are systems that are built by combining a *CI/AI* method that is inadequate with respect to one key *HIS* feature with a method that is good to excellent with respect to that key feature. A lot of function-replacing *HIS* is already traditional. See for example: *GA-NN HIS* that are used for *NN* optimisation (*NN* weights and/or structure adjustments); *GA-FS HIS* that are used for optimising both the rules structure (number of fuzzy labels) and the *FS* parameters (fuzzy shape, position of fuzzy variables).

Intercommunicating HIS are systems relying on *technique-to-task allocation*: a complex application problem is sub-divided in specialised component problems (tasks), and a suitable *CI/AI* technique is allocated to each of these tasks. So different component problems, each of which may require different type of information processing, are solved by different specific *CI/AI* techniques and the final (global) results of the *HIS* are performed by communicating the specific results to each *CI/AI* techniques among themselves. See for example what is involved in the problem of launching a new manufactured product: a *forecasting task* is solved by using a *NN* technique, a *multi-objective optimisation* task is solved by using *GA* techniques and may be other reasoning tasks would use *FS/KBES* techniques.

The first kind of *HIS* classification relies on real-world application requirements and groups the hybrid systems into two as mentioned above. An engineering point of view rather than a "systemic" one might be kept even if we are looking to theoretical (structural) aspects of information processing in *HIS*. In this case *HIS* are grouped into four classes: *fusion* systems, *transformation* systems, *combination* systems and *associative* systems (Khosla and Dillon 1997).

Fusion HIS is featured by a new *CI/AI* technique, *X*, whose procedure of information processing and/or its representation features are melted into the representation structure of a host *CI/AI* technique, Y. The practical aspects of melting a new *CI/AI* technique in a fusion *HIS* means that the host technique Y is diminishing its own weakness and exploits its strength more effective for solving the real-world application under these new circumstances. Let us mention some applications of *NN* based hybrid systems in which *FS* information processing features are fused: fuzzification of input data, optimisation of fuzzy systems, modelling fuzzy inference in *NN* structures, as well as implementation of *FS* logic operations in *NN*. Another familiar example of a fusion hybrid system is when *NN* adaptive properties are used to optimise the performance of a fuzzy rules-based controller. The fusion of fuzzy rules and membership degrees into a *NN* is made. Then the reinforcement learning or backpropagation algorithms are used to optimise the performance. It does this by changing the shapes of the membership functions, fine tuning these shapes through the adaptation of *NN* weights or through the changing of weight links. A well known *GA* based hybrid system in which *FS* information processing features are fused is the Lee/Takagi *GA* based system of intelligent integrated *FS* design. In this system both the membership functions and the number of useful fuzzy rules, as well as rule-consequent parameters are designed automatically and optimised using *GA*. See (Lee and Takagi 1993).

Transformation HIS are featured by the transformation of one form of information representation into another one. These hybrid systems are used for applications where the required knowledge for task achievement is not available and where one *CI/AI* method depends upon a different *CI/AI* method for its reasoning and processing purposes. Let us mention some example of transformation *HIS*. Two typical applications solved by neuro-fuzzy transformation systems are now described. The first is the use of learning fuzzy *IF-THEN* rules, learning fuzzy clusters and learning membership functions. Here *NN* learning features are integrated as a method operating on input-output data pairs or simply on input data. The second type of transformation *HIS* is the NN-*KBES* systems that are performing concept hierarchies or *KBES* rules from a *NN*.

A special case of transformation *HIS* occurs in special application circumstances where transformation is required to be in form of a complex optimisation. Firstly, an un-optimised technique X is melted into representation Y. Secondly, an extraction of the optimised prior representation X is performed from optimised representation Y. This is an intermediated class of *HIS* that is usually called a *fusion and transformation HIS*. Such systems are *GA-NN* or *GA-FS-NN* systems performing a complex GA optimisation of *NN* or *fuzzy NN*, namely envisaging the *NN* connection weights, *NN* structure, *NN* topology and *NN* input data. See (Arotaritei and Negoita 2002; Palade et al. 1998). During the last decade, all *CI/AI* methods have got an increased audience on behalf of practitioners, so these techniques have been involved in a large variety of real-world applications. The approach of using these technologies by themselves alone has exposed real limitations. These limitations are nothing else than the result of the complexity associated with problems from the real-world work environment. There has also developed another class of hybrid system called *intelligent combination HIS* in which two or more *CI/AI* technologies are combined. The distinct role played by each of the technologies in combination results in a more effective problem solving strategy. This combination *HIS* approach has an explicit hybridisation essence that relies on a refined matching of the *CI/AI* methods to the particular components of a modular structured model. The most common combination hybrid systems are now described. *FS-NN-KBES* complex (fault) diagnosis systems in industry and medicine. *FS-NN-GA* hybrid systems for developing intelligent *GA* learning based adaptive control applications with *FS-NN* controllers. Combination systems performing a large variety of *KDD* applications in the finance and banking areas. (*KDD* stands for **K**nowledge **D***iscovery* in **D***atabases*). Then there are general purpose *GA-KBES* combination scheduling systems. General purpose *FS-GA-KBES* customer advisor systems advising people how and what to buy. Combination systems performing a large variety of *KDD* specialised forecast/prediction tasks and also combination systems for complex tasks in robotics.

The four *HIS* classes, namely *fusion HIS, transformation HIS, fusion-transformation HIS* and *combination HIS* are motivated deeply by real-world applications. From the point of view of the quality of task achieved and the range of task covered when applications have been suitably matched to the available technologies, development in all these *HIS* categories has led to spectacular results.

However it must be noted that no *HIS* developed within these categories can be classified as being a general purpose technology – capable of handling any application. All four classes of *HIS* have their limitations. A typical drawback, particularly with the fusion and transformation systems, is that they sometimes are not able to capture all aspects of human cognition related to solving the application. Even the combination *HIS* which seems to be the most complete system, expresses a lack of minimal knowledge transfer among modules, despite their system flexibility. The range of tasks covered by fusion systems is restricted by loosing of the declarative aspects of solving the application. This is due to conversion of explicit knowledge into implicit knowledge.

The drawbacks of fusion, transformation, fusion-transformation and combination systems favours the implementation of a more powerful system in solving applications, called *associative HIS*. Associative *HIS* may incorporate fusion, transformation and combination architectural/concept elements as well as integrate standalone *CI/AI* techniques. The results of these strategies confer a much-improved *HIS* quality features with respect to the quality of task achieved and to the range of tasks covered. See (Khosla and Dillon 1997) for more details on associative *HIS* as well as for a wonderful description of a lot of *HIS* applications.

As a conclusion, hybridisation meets application requirements and is strongly related to real-world problems. Hybridisation might be considered as an hierarchical process. The first hybridisation level caters to applications of a low to moderate complexity and consists of just combining some *CI/AI* techniques to form an *HIS*. The second hybridisation level is typical to applications of high complexity where some *HIS* is combined, together or not, with some stand alone *CI/AI* techniques. If we consider just the "silicon environment", these two levels of hybridisation are nothing more than an evolution of design strategies. However, it is now possible to look past simply the "silicon environment". Despite still being very much in the pioneering stage, quite promising steps have been made in a possible transition from the silicon environment hosted by microchips to the so-called "carbon environment" hosted by *DNA* molecules. Organic molecules containing *DNA* molecules intrinsically provide huge results of information processing capability. So the way is paved for a third hybridisation level performing hybridisation of some stand alone *CI/AI* techniques, even of some complete *HISs*, along with the biological advances being made. The third level of hybridisation means a bio-molecular implementation of soft computing, so that the uncertain and inexact nature of chemical reactions inspiring *DNA* computation will lead to implementation of a new generation of *HIS*. The so-called **R**obust (**S**oft **C**omputing) **H**ybrid **I**ntelligent **S**ystems – *RHIS*.

RHIS are systems of biological intelligence radically different from any kind of previous intelligent system. The difference is expressed in three main features:

- *robustness* – conferred by the "carbon" technological environment hosting the *RHIS*
- *miniaturization* of the technological components at a molecular level
- the highest (*biological*) *intelligence* level possible to be implemented in non living systems
- dealing with world knowledge in a manner of *high similarity to human beings*, mainly as the result of embedding *FL*-based methods of **C**omputing with **W**ords and **P**erceptions (*CWP*) featured by the understanding that perceptions are described in a natural language.

In the few years, interest has been growing in the use of a large variety of *HIS*. At every level of hybridisation, the use of biologically inspired *CI* techniques has been decisive in improving *HIS* features and performances. An emerging and promising biologically inspired technique is the field of **A**rtificial **I**mmune **S**ystems (*AIS*). These are adaptive systems inspired by theoretical immunology, especially the vertebrate immune system and by observed immune functions and principles. These *AIS* are applied to typical problems previously being solved by *HIS* such as: pattern recognition, data analysis (clustering), function approximation and multi-modal optimisation (de Castro and Timmis 2002). Results got in some applications regarding function optimisation show that *AIS* algorithms are more effective than *GA*. *AIS* perform significantly fewer evaluations than a *GA* without altering the quality of the final solution (Kelsey and Timmis 2003).

2 Neuro-Fuzzy Based Hybrid Intelligent Systems for Fault Diagnosis

In the last ten years, the field of diagnosis has attracted the attention of many researchers, both from the technical area as well as from the medical area. In the industrial field there is also an increasing need for reliability, which resulted in the development of various techniques for an automatic diagnosis of faults. Generally, in an industrial control system a fault may occur as follows: in the process components, in the control loop (controller and actuators) and in the measurement sensors for the input and output variables. The conceptual scheme for a fault diagnosis system consists of two sequential steps: residual generation and residual evaluation. In the first step, a number of residual signals are generated in order to determine the state of the process. The objective of fault isolation is to determine if a fault has occurred and, if so, to determine the location of the fault, by an analysis on the residual vector.

2.1 Introduction

Many authors have focused on the use of *NNs* in **F**ault **D**etection *and* **I**solation *(FDI)* applications (Marcu et al. 1999; Korbicz et al. 1999) for solving the specific tasks in *FDI*, mainly fault detection but also fault isola-tion. Other authors (Koscielny et al. 1999) used *FSs* for fault diagnosis and especially for fault isolation. Some authors even used *FSs* for fault detection, using for example *TSK* fuzzy models. In the last few years there has also been an increasing number of authors (Patton et al. 1999; Calado and Sa da Costa 1999) who have tried to integrate *NNs* and *FSs* in the form of *NN-FS HIS*, in order to benefit of the advantages of both techniques for fault diagnosis applications.

    NNs have been successfully applied to fault diagnosis problems due to their capabilities to cope with non-linearity, complexity, uncertainty, noisy or corrupted data. *NNs* are very good modeling tools for highly non-linear processes. Generally, it is easier to develop a non-linear *NN* based model for a range of operating tasks, than to develop many linear models, each one for a particular operating point. Due to these modeling abilities, *NNs* are ideal tools for generating residuals. *NNs* can also be seen as universal ap-proximators. A common *3* layered *MLP NN*, with *m* inputs *and n* outputs,

Mircea Gh. Negoita, Daniel Neagu, and Vasile Palade: *Computational Intelligence: Engineering of Hybrid Systems*, StudFuzz **174**, 13–24 (2005)
www.springerlink.com
© Springer-Verlag Berlin Heidelberg 2005

can approximate any non-linear mapping from R^m to R^n using an appropriate number of neurons in the hidden layer. Due to this approximation and classification ability, *NNs* can also be successfully used for fault evaluation. The drawback of using *NNs* for classification of faults is their lack of transparency in human understandable terms. *FS* techniques are more appropriate for fault isolation because they allow for a more natural integration of human operator knowledge into the fault diagnosis process. The formulation of the decisions taken for fault isolation is done in a human understandable form, such as linguistic rules.

The main drawback of *NNs* is represented by their "black box" nature, while the disadvantage of *FSs* is in the difficult and time-consuming process of knowledge acquisition. On the other hand, the advantage of *NNs* over *FSs* is its learning and adaptation capabilities, while the advantage of *FSs* is the human understandable form of knowledge representation. *NNs* use an implicit way of knowledge representation, whilst *FSs* and *NN-FS HISs* represent knowledge in an explicit form, such as rules.

2.2 Methods of NN-FS Hybridization in Fault Diagnosis

The *NN-FS* hybridization can be done in two main ways:

(a) *NNs is the basic methodology and FSs is the second.* These *NN-FS HISs* exhibit mainly *NN* features, but the *NNs* are equipped with the ability to process fuzzy information. These *HISs* are also termed *Fuzzy Neural Networks* (*FNN*) and they are systems where the inputs and/or the outputs and/or the weights are fuzzy sets. They are usually built using a special type of neuron, called fuzzy neuron. Fuzzy neurons are neurons with inputs and/or outputs and weights represented by fuzzy sets, the operation performed by the neuron being a fuzzy operation.

(b) *FSs is the basic methodology and NNs the subsequent.* These *HIS* systems can be viewed as *FSs* augmented with *NNs* facilities, such as learning, adaptation, and parallelism. These *HIS* systems are also called *Neuro-Fuzzy Systems* (*NFS*). Most authors in the field of neuro-fuzzy computation understand *neuro-fuzzy systems* as a special way to learn *FSs* from data using *NN* type learning algorithms. Some authors (Shann and Fu 1995) term these *neuro-fuzzy systems* also as *fuzzy neural networks*, but most of the authors in the are like to refer to them as *neuro-fuzzy systems*. NFS-type HISs can always be interpreted as a set of fuzzy rules and can be represented as a feed-forward network architecture.

These two ways of *NN-FS* hybridization presented above can be viewed as a type of *fusion HIS*, as it is difficult to see a clear separation between the two methodologies (Khosla and Dillon 1997). One methodology is fused into the other methodology, and it is assumed that one technique is the basic technique and the other is fused into it and augments the capabilities of information processing of the first methodology. Besides these *fusion NN-FS HIS* systems, there is another way of hybridization of *NNs* and *FSs*, where each methodology maintains its own identity and the hybrid *NN-FS* system consists of a modules structure which cooperate in solving the problem (Khosla and Dillon 1997). These kind of *NN-FS HISs* are called *combination HIS* systems. The *NN*-based modules can work in parallel or serial configuration with *FS*-based modules and augment each other. In some approaches, a *NN*-such as a self-organizing map- can preprocess input data for a *FS*, performing for example data clustering or filtering noise. But, especially in *FDI* applications, many authors use a *FS* as a pre-processor for a *NN*. In (Alexandru et al. (2000)) the residual signals are fuzzified first and then fed into a recurrent *NN* for evaluation, in order to perform fault isolation.

The most often used *NN-FS HIS* systems are *fusion neuro-fuzzy systems*, and the most common understanding for such a system is the following. A (fusion) *neuro-fuzzy system* is a *NN* which is topologically equivalent to the structure of a *FS*. The network inputs/outputs and weights are real numbers, but the network nodes implement operations specific to *FSs*: fuzzification, fuzzy operators (conjunction, disjunction) and defuzzification. In other words, a *NFS* can be viewed as a FS, with its operations implemented in a parallel manner by a *NN*. That's why it is easy to establish a one-to-one correspondence between the *NN* and the equivalent *FS*. (Fusion) *NFSs* can be used to identify fuzzy models directly from input-output relationships, but they can also be used to optimize (refine/tune) an initial fuzzy model acquired from human expert, using additional data. Two most often used systems of this type are presented in the next section.

2.2.1 Neuro-Fuzzy Networks

In the area of *NFSs* there are two principal types of *neuro-fuzzy networks* (*NFN*) preferred by most of the authors in the field of *NN-FS* hybridization. The most common *NFN* is used to develop or adjust a fuzzy model in Mamdani form, given by the following relation, using input – output data. The *HIS* is structured in the form of a five layers network as shown in Fig. 2.1. A Mamdani fuzzy model consists of a set of fuzzy *IF-THEN* rules in the following form:

$$\text{IF } x_1 \text{ is } X_{1i_1} \text{ and } x_2 \text{ is } X_{2i_2} \text{ and} \ldots \ldots x_n \text{ is } X_{ni_n}$$
$$\text{THEN } y \text{ is } Y_j \tag{2.1}$$

where:
$x_1 x_2, \ldots, x_n$ are the system inputs,
y is the output,
X_{ki_k} with $k = 1, 2, \ldots, n$ and $i_k = 1, 2, \ldots, l_k$ are the linguistic values of the linguistic variable x_k,
$Y_j, j = 1, 2, \ldots l_y$, are the linguistic values of the output variable.

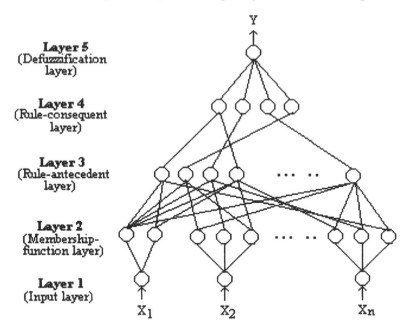

Fig. 2.1. The general structure of a *NN-FS HIS* network for Mamdani models

Every linguistic variable x_k is described by l_k linguistic values X_{k1}, X_{k2}, $\ldots, X_{kl_k}$.

Layer 1 is the input layer and each node in this layer corresponds to an input variable.

Layer 2 is called membership function layer, the nodes from this layer mapping each input x_i to every membership function X_{ij} of the linguistic values of that input. It is possible to use, in layer 2, a subnet of nodes to implement a desired membership function, instead of a single node.

Each node in layer 3 (called rule layer) performs the precondition matching – the *IF* part – of a fuzzy rule.

The nodes from layer 4 combine the fuzzy rules with the same consequent, each node implementing a fuzzy *OR* operator, such as fuzzy *MAX* operator.

Each node in layer 5 corresponds to an output variable and acts as a defuzzifier.

The integration and the activation functions of nodes for such an *HIS* network are chosen (Shann and Fu 1995) so that they perform specific operations in a fuzzy inference engine as described before.

Another major class of *NN-FS HIS* network structure is the *NFN* used to develop and adjust a Sugeno-type fuzzy model. The structure of such a *neuro-fuzzy network* is shown in Fig. 2.2.

The first 3 layers are the same as those in a *NFN* for Mamdani models. In the rule layer, the traditional fuzzy *MIN* operator can be used, but many authors prefer to use a PRODUCT operator, as a fuzzy intersection operator.

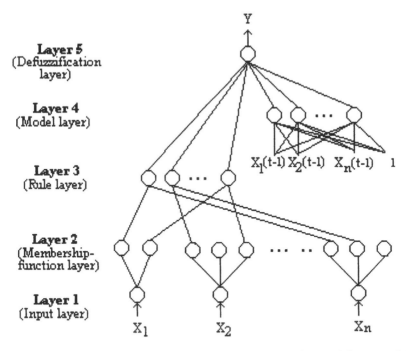

Fig. 2.2. The general structure of a *NN-FS HIS* network for TSK fuzzy models

Usually, all weights in this layer are set to 1. If some prior knowledge on process functioning is available, we can set up the number of nodes in layer 3 (the number of rules or fuzzy partition regions) and the corresponding links between layer 2 and 3.

A *NN-FS HIS* network structure was developed in (Zhang and Morris 1996) for process modeling and fault diagnosis. The main shortcoming of this structure is that the user must partition the process operation into several fuzzy operating regions before training the *NN-FS* network. The partitioning is made empirically, looking at the process functioning, and it may be a very difficult task when the process has a complex nature. Different clustering techniques, as well as *GAs* (resulting in a *NN-FS-GA HIS* optimized network structure), can be used to find the best fuzzy partition of the input space.

Layer 4 is called the model layer, and each node implements a linear model corresponding to a rule node in the rule layer, respectively to a fuzzy operating region. The weights of a node are the parameters of the linear model and the inputs of the node are the past system inputs and outputs.

Layer 5 consists of a single node that performs the defuzzification.

The most general Sugeno-type *NN-FS HIS* network structure is a network which implements a set of fuzzy rules with *ARMA* models of higher order in the consequence part of the rules. When linear *ARMA* models of higher order are used, every node from layer 4 must be replaced by a subnet (see Fig 2.3),

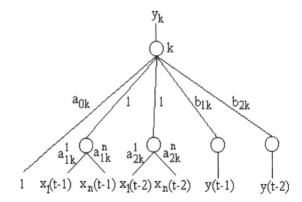

Fig. 2.3. The subnet corresponding to node k in layer 4

which implements the *ARMA* model of the desired order. The rules are in the following form:

R_k: IF x_1 is X_{1i_1} and x_2 is X_{2i_2} and...x_n is X_{ni_n} THEN

$$y_k(t) = a_{0k} + \sum_{j=1}^{n_1} a_{jk} x(t-j) + \sum_{j=1}^{n_2} b_{jk} y(t-j) \qquad (2.2)$$

where $k = 1, 2, \ldots, m$, m the number of rules, and $x = (x_1, x_2, \ldots, x_n)$ is the input vector, and $a_{jk} = (a_{jk}^1, \ldots, a_{jk}^n)$.

In Fig. 2.3, it is shown the subnet which corresponds to node k from layer 4, when $n_1 = n_2 = 2$. The inputs of the subnet k from layer 4 are the previous inputs and outputs of the system.

2.2.2 Residual Generation Using Neuro-Fuzzy Models

The purpose of this section is to present a *NN-FS HIS* application to detect and isolate actuator faults mainly, but also other faults such as components or sensor faults that occur in an industrial gas turbine. In this turbine model, air flows via an inlet duct to the compressor and the high pressure air from the compressor is heated in combustion chambers and expands through a single stage compressor turbine. A *Butterfly valve* provides a means of generating a back pressure on the compressor turbine (there is no power turbine present in the model). Cooling air is bled from the compressor outlet to cool the turbine stator and rotor.

A *governor* regulates the combustor fuel flow to maintain the compressor speed at a set-point value. For simulation purposes we used a *Simulink* prototype model of such an industrial gas turbine as presented in (Patton and Simani 1999) and developed at ABB-Alstom Power, United Kingdom. The *Simulink* prototype simulates the real measurements taken from the gas turbine with a sampling rate of 0.08 s. The model has two inputs and 28 output measurements which can be used for generating residuals. The *Simulink*

model was validated in steady state conditions against the real measurements
and all the model variables were found to be within 5% accuracy. Different
NN-FS HIS models were developed for generating residuals purposes, but all
of them are driven by two inputs: valve angle (*va*) and the fuel flow (*ff*),
which is also a control variable, see Fig. 2.4.

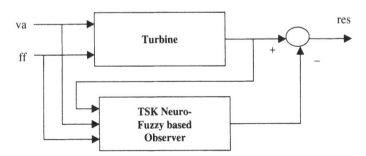

Fig. 2.4. *NN-FS HIS* based observer scheme for generating residuals

One common fault in the gas turbine is the fuel actuator friction wear
fault. Other faults considered were the compressor contamination fault, ther-
mocouple sensor fault and high-pressure turbine seal damage (Palade et al.
2002). Usually these faults in the industrial gas turbine develop slowly over
of a long period of time. During the simulations, an attempt was made to
detect the actuator fault and to isolate it from other faults in the turbine.
For simplicity reasons, this section presents only the results for the first two
faults – fuel actuator friction wear fault and compressor contamination fault.

The residual signals are given by the difference between the estimated
signal given by observer and the actual value of the signal. The residuals are
generated using *TSK neuro-fuzzy networks*. A main concern was focused on
how to find accurate *NN-FS HIS* models for generating residuals, but which
still have as much transparency as possible in the models. That's why a good
structure had to be found for the *NN-FS HIS* models and, therefore, a good
partition of the input space, using clustering techniques. There is always a
compromise between the interpretability and the precision of the model.

First, a 3 inputs and 1 output *TSK* neuro-fuzzy network structure was
developed for the output measurement which is most affected by the actuator
fault (*ao*). The inputs of the network are the present value for valve angle
(*va*) and fuel flow (*ff*), and the previous value of the output affected by the
fault. Three linguistic values were used for each input variable as well as
grid partition of the input space. See (Palade et al. 2002) for more details
regarding the performance of the model, the generated residuals and the
difference between the system output and the model output. Unfortunately,
due to control loop action, this kind of fault can not be seen in steady state
regime and can not be described as a ramp function in order to show what
happened in case of a gradual developing fault. But this actuator fault can

be seen in a dynamic regime, for different values of the valve angle. For isolating purposes the absolute value of this residual signal was taken and passed through a filter for obtaining a persistent residual signal.

In order to see a gradually developing fault, a *TSK neuro-fuzzy* based observer was built for the output most affected by the compressor contamination fault. In a similar way a *TSK NFN* with 3 inputs and 3 membership functions per input and first order linear models in the consequent of the rules was used. The output (*co*), most sensitive to a compressor ramp fault, was depicted and the residual generated too in (Palade et al. 2002). The compressor fault also affected the output *ao*, but not in the same magnitude as the output *co*. Then, the residual designed for output *ao* is sensitive to both faults. See (Palade et al. 2002) for more details.

Several other types of models were developed for the residual sensitive to the actuator fault, in order to see a comparison between the accuracy and the transparency of the model. These results are summarized in Table 2.1. The results in this table led to the concluding remark that if a more transparent *TSK neuro-fuzzy model* is needed for residual generation, the accuracy of the model will be gradually lost. The first three *TSK* models were generated using clustering methods and the following three were generated using a grid partition with 2, 3 and 4 membership functions for each input variable. The exceptional performance shown in the first case can be explained by the capabilities of *Gustafson-Kessel* clustering method, which can produce clusters with different shapes and orientation, that result in a more accurate model, while keeping a reduced number of rules.

Table 2.1. Performance evaluation

Transparency of the Model (no. of rules)	Performance of the Model	Comments
2 rules	0.00180	Gustafson-Kessel
2 rules	0.007214	Substr. Clustering
3 rules	0.006087	Substr. Clustering
8 rules ($2 \times 2 \times 2$)	0.004194	Grid partition
27 rules ($3 \times 3 \times 3$)	0.001024	Grid partition
64 rules ($4 \times 4 \times 4$)	0.000887	Grid partition
Black box	0.000001	Neural network

On the other hand, it can be concluded that if the structure of the *TSK* model is not properly chosen, then not only is the degree of transparency reduced, but the training time will be greatly increased, and it will be far more difficult to reach a desired performance of the model in a given time.

Table 2.1 also shows the concluding remark that there is not a very big difference in performance between a model with 64 rules and a model with 2 rules, even if the number of the parameters were much bigger in the first case. That means the structure of the model in the first case was not appropriate and the training sessions were not fully completed. In fact, after an appropriate and complete training, the model with 64 rules should have overlapping membership functions and many input regions with about the same consequent, which require a post-processing of the model in order to simply it. But this task, represented by a more difficult training followed by a simplification of the model, is more complicate to perform than trying to predict first a right structure of the model at a desired degree of transparency and train the *NN-FS HIS* model after that.

2.2.3 Neuro-Fuzzy Based Residual Evaluation

In the residual generation part of a diagnosis system the user should be more concerned with the accuracy of *NN-FS HIS* models, even desirable to have interpretable models also for residual generation, such as *TSK NN-FS HIS* models. For the evaluation part, the transparency or the interpretability of the fault classifier in human understandable terms, such as classification rules, is more important.

The main problem in *NN-FS HIS* fault classification is how to obtain an interpretable fuzzy classifier, which should have few meaningful fuzzy rules with few meaningful linguistic rules for input/output variables. *NN-FS HIS* network structures for Mamdani fuzzy models are appropriate tools to evaluate residuals and perform fault isolation, as the consequence of the rules contains linguistic values which are more readable than linear models in case of using *TSK* fuzzy models. As mentioned in the previous section, a price was paid for the interpretability of the fault classifier, that is the loss of precision of the classification task.

More details regarding the experiments and their concluding remarks on residual evaluation can be found in (Palade et al. 2002). For training the *NN-FS HIS* networks (*Mamdani NFNs*) in order to isolate these faults, 150 patterns for each fault were used. The *NN-FS HIS* network decisions for the residual values were assigned in relation with the known faulty behavior. In order to obtain a readable fault classifier, the NEFCLASS neuro-fuzzy classifier was used (Nauck and Kruse 1998). NEFCLASS has a slightly different structure than the *NFN* for Mamdani models presented in the previous section, but it allows to the user to obtain, in an easy and interactive manner, an interpretable *FS* fault classifier, at the desired level of accuracy and transparency. Conjunctive fuzzy rules for fault classification (using both residual inputs in the antecedent) were considered.

Table 2.2 summarizes the results of the simulation experiments (Palade et al. 2002) featuring the transparency/accuracy compromise of the *NN-FS HIS* fault classifiers.

Table 2.2. Transparency/Accuracy of the evaluation task

Transparency (no. of rules)	2	4	8	12
Accuracy (no. of patterns correctly classified in %)	88.7	91.2	96.4	99.6

2.3 Case Example: Fault Detection and Isolation of an Electro-Pneumatic Valve

Mathematical models used in the traditional **F**ault **D**etection and I*solation* (*FDI*) methods are sensitive to modeling errors, parameter variation, noise and disturbance (Uppal et al. 2002). In addition to this, pocess modeling has severe limitations when the system is complex and uncertain, or when the data is noisy or incomplete. **C***omputational* **I***ntelligence* (*CI*) methods – **N***eural* **N***etworks* (*NN*), **F***uzzy* **S***ystems* (*FS*), **E***volutionary* **A***lgorithms* (*EA*) are able to overcome some of the above mentioned problems (Patton et al. 2000).

The *NN-FS HIS* models combine both numerical and symbolic knowledge about the process. Automatic linguistic rule extraction is a strong point in favor of *NN-FS HISs*, especially when no prior knowledge about the process is available. A *NN-FS HIS* model can be identified directly from empirical data. These neuro-fuzzy models identified from empirical data are usually not very transparent. In case of complex systems, hierarchical *NN-FS HIS* networks can be used to overcome the dimensionality problem by decomposing the system into a hierarchical structure of *MISO* systems (Tachibana and Furuhashi 1994).

The criteria on which to build *NN-FS HIS* models for *FDI* applications (Uppal et al. 2002) should consider the function of the *NN-FS HIS* model in the *FDI* scheme, i.e.: Preprocessing data, Identification (Residual generation) or Classification (Decision Making/Fault Isolation). A *NN-FS HIS* model with high approximation capability and disturbance rejection is needed for residual generation. In the classification stage, a *NN-FS HIS* network with more transparency is required.

The valve considered for *FDI* in this case example (Uppal et al. 2002) is an electro-pneumatic flow control valve used in the evaporisation stage of a sugar factory. A non-linear mathematical model of the valve was developed in *Matlab – Simulink*. This model was then used to generate faulty and fault-free data in order to evaluate different *NN-FS HIS* based fault isolation schemes. The valve assembly consists of 3 main parts, as in Fig. 2.5:

The PI controller – which controls the *Positioner* and the *Actuator* output in order to regulate the juice flow through the valve.

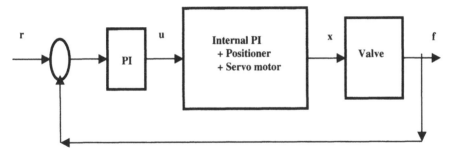

Fig. 2.5. Main parts of the valve assembly

Positioner and Actuator – Pneumatic pressure is applied to the servomotor diaphragm to control the stem position that changes the flow. The *Positioner* adjusts this pressure input to the servomotor to obtain the correct stem position of the *actuator*.
The Valve – which alters the flow according to the stem position.

The following list of faults was considered in the valve actuator assembly (Uppal et al. 2002):

f_1 – External *PI* controller proportional gain fault
f_2 – External *PI* controller integral gain fault
f_3 – Increased friction of the servomotor
f_4 – Decreased elasticity of the servomotor
f_5 – Decrease of pneumatic pressure
f_6 – Internal *PI* controller fault
f_7 – Internal position sensor fault
f_8 – Valve clogging
f_9 – Valve leakage
f_{10} – Chocked flow

Two *NN-FS HIS* models with a transparent structure were used. A *TSK* neuro-fuzzy structure with linear dynamic models in the consequent part of the rules was used to approximate the internal *PI* controller, the *Positioner* and *Servomotor*. The identified *TSK* model had three local linear models in the consequent of the rules. More details regarding the performance of the *TSK* models in closed-loop are to be found in (Uppal et al. 2002).

The changes in physical parameters, e.g. time constant (r_{TC}), static gain (r_{SG}), static offset (r_{SO}) and settling time (r_{ST}) are computed using local models and can be used for fault isolation.

A linguistic Mamdani *NN-FS HIS* model was identified for fault isolation. The model input is the stem position x and the output is the volumetric flow rate f. The control input u is predicted from the input set-point flow and measured flow. *GK*-clustering algorithm (Gustafson and Kessel 1979) was used to partition the input space. The clusters were projected onto the input/output

Table 2.3. Fault isolation for valve assembly

	f_1	f_2	f_3	f_4	f_5	f_6	f_7	f_8	f_9	f_{10}
r_u	Op + Cl −	Op − Cl +	~	~	~	~	~	~	~	~
r_x	~	~	~	~	~	~	Ch	~	~	~
r_f	~	~	~	~	~	~	~	Op + Cl −	Op − Cl +	−
r_{ST}	~	~	+	+	0	−	~	~	~	~
r_{TC}	~	~	+	−	+	−	~	~	~	~
r_{S0}	~	~	0	0	0	+	~	~	~	~

Op+:   Positive value when valve is being opened

Op−:   Negative value when valve is being opened

Cl+:   Positive value when valve is being closed

Cl−:   Negative value when valve is being closed

Ch: Changed

space to find membership functions. A gradient-based optimization method was used to fine-tune the membership functions. More details on both clustered valve data and on the tuned membership functions for Mamdani model are to be found in (Uppal et al. 2002). The predicted values u, x, f and the measured values were used to generate the residuals r_u, r_x, r_f. Fault isolation table (Uppal et al. 2002) given in Table 2.3 shows that some faults can be detected only when the valve is opened or closed. Choked flow could only be detected at high values of flow.

3 Neuro-Fuzzy Integration
in Hybrid Intelligent Systems

In the last fifteen years, hybrid neural systems have drawn increasing research interest. This approach has been successfully used in various areas, such as speech/natural language understanding, robotics, medical diagnosis, fault diagnosis of industrial equipment, and financial applications (Kosko 1992; Lin and Lee 1991; Pedrycz 1993; da Rocha 1992; Sima and Cervenka 1997; Takagi 1994; Wermter and Sun 2000). The reason for studying hybrid neural systems is based on successful applications of subsymbolic knowledge representation systems, particularly the ones based on the neuro-fuzzy networks paradigm, as well as the advantages of the symbolic knowledge-based systems. From the point of view of cognitive science, the purely neural representation offers the advantage of homogeneity, as well as the possibility of performing distributed tasks, and working with incomplete and noisy data. From the point of view of knowledge-based systems, symbolic representations have advantages of human interpretation, explicit control, and knowledge abstraction (Kosko 1992; Wermter and Sun 2000).

The two approaches used for the development of neural expert systems are knowledge extraction from trained neural networks, and the integration of implicit (connectionist) knowledge with symbolic expert systems (Hilario 1997; Sima and Cervenka 1997). In these hybrid systems, connectionist tools can be interpreted as hardware, and fuzzy logic as software implementation of human reasoning. In that way, the modular structure of connectionist implementations of explicit and implicit knowledge can be interpreted as a homogenous system combining inductive and deductive learning and reasoning.

This chapter proposes a unified approach for integrating implicit and explicit knowledge in neurosymbolic systems as a combination of neural and neuro-fuzzy modules. In such developed hybrid systems the training data set is used for building neuro-fuzzy modules, and represents implicit domain knowledge. The explicit domain knowledge on the other hand is represented by fuzzy rules, which are directly mapped into equivalent neural structures. The aim of this approach is to improve the abilities of modular neural structures, which are based on incomplete learning data sets, since the knowledge acquired from human experts is taken into account for adapting the general neural architecture.

Mircea Gh. Negoita, Daniel Neagu, and Vasile Palade: *Computational Intelligence: Engineering of Hybrid Systems*, StudFuzz **174**, 25–39 (2005)
www.springerlink.com
© Springer-Verlag Berlin Heidelberg 2005

A brief overview of the motivations and the state of the art in the area of neurosymbolic integration will be presented in the next section. The fundamental concepts and methods used in our approach are described subsequently: the formalism of the neural fuzzy model MAPI (introduced by da Rocha in 1991); the concept of implicit knowledge and proposed specific architectures used to implement it; the concept of explicit knowledge and two connectionist methods proposed to implement a given fuzzy rule set.

The implicit knowledge is defined as a connectionist module-based representation of learning data. Subsequently, the explicit knowledge module of the hybrid system is implemented as a special connectionist structure using hybrid fuzzy neural networks. This kind of implementation is proposed to adjust the performances of implicit knowledge modules.

Of course, there are various approaches to represent knowledge following the definition of explicit and implicit knowledge. One might find similarities with the process to define, develop and represent in a connectionist manner information about the concept "animal" and some particular entities in an extrinsic distributed representation (as depicted in Fig. 3.1) or, for the same concepts, in an intrinsic distributed representation (see Fig. 3.2).

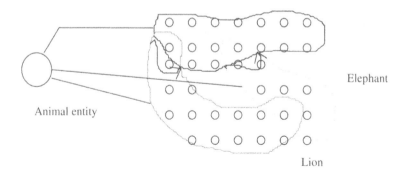

Fig. 3.1. Extrinsic Distributed Knowledge Representation

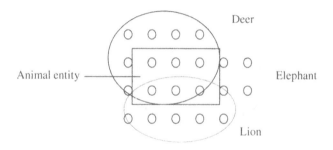

Fig. 3.2. Intrinsic Distributed Knowledge Representation

3.1 Neuro-Fuzzy Integration

The last ten years have produced an explosion in the amount of research on both symbolic and connectionist fields. Symbolic processing is considered a traditional way in Artificial Intelligence. In connectionist systems, unlike symbolic models, learning plays a central role. A combination of both approaches is already the subject of research in hybrid systems. Connectionist models are powerful tools to process knowledge. In this way they have been used to build connectionist intelligent systems, mainly for perceptual tasks, where discovering explicit rules does not seem either natural, or direct. Like human beings, the connectionist models rely on learning low-level tasks. Learning by example is not a general solution: it is known that many situations are solved by intelligent entities using explicit rules.

The two approaches can be used in a complementary way. This is the premise of hybrid intelligent systems which combine connectionist and symbolic features. In such systems, since the learner must use prior knowledge in order to perform well, the learner first inserts symbolic information of some sort into the neural network. Once the domain knowledge is put in a neural representation, training examples are used to refine the initial knowledge. Finally, it processes the output for a given instance and, using some specific methods (Benitez et al. 1996, 1997; Neagu and Palade 2000a,b; Palade 1999) extracts symbolic information from the trained network in order to give some explanations on the computed output and improve understanding of the refined connectionist knowledge. Since these steps are somewhat independent, it is evident that building hybrid intelligent systems requires an exploration of all these approaches. The most important effort in such a situation is focused to the homogenous implementation of all these methods into connectionist structures, in order to use their capabilities to perform well and provide accurate conclusions using often, incomplete and noisy data. In this work it is proposed to represent some approaches in a neural manner, namely external fuzzy rules explicitly acquired from human experts. These rules are combined with trained neural structures, built using data sets of the same application domain, in order to improve accuracy of the output.

Complex tasks may give rise to local minima; subsequently the "learning by example" paradigm is useful mostly in simple tasks. Our proposed answer to this problem is to combine different connectionist modules, solving various subtasks of the main problem. The connectionist integration of explicit knowledge and learning by example appears to be a natural solution of developing connectionist intelligent systems. The problem to be solved is the uniformity of integration. In order to encourage modularization, explicit and implicit rules should be represented in a neural manner using (Buckley and Hayashi 1995; Fuller 1999) fuzzy neural networks (FNN), hybrid neural networks (HNN), and, in a particular approach, standard neural networks (multilayer perceptron-based structures MLP, Rumelhart and McClelland 1986). While fuzzy logic provides the inference mechanism under cognitive uncertainty, neural networks offer the advantages of learning, adaptation, fault-tolerance, parallelism and generalization (Fuller 1999).

The computational process involved in implicit and explicit knowledge acquisition and representation is described in the following sections. Firstly, the definition of a "fuzzy neuron" is provided, based on the understanding of biological neuronal structure, followed by learning mechanisms (for implicit knowledge representation as combined standard and/or fuzzy neural networks), respective fuzzy rule mapping mechanisms (for explicit knowledge representation as hybrid neural networks). This leads to the three steps in a neural computational process (Fuller 1999): development of fuzzy neural models; models of synaptic connections, which incorporate fuzziness into neural network; and application of learning, respective mapping algorithms to adjust the synaptic weights. The system taken into consideration in the next sections is a multi-input single-output fuzzy system (MISO).

3.1.1 The Fuzzy Neuronal Model

The MAPI neuron, proposed by Rocha (da Rocha 1991, 1992; Pedrycz and da Rocha 1993), is a useful tool to add another level of programmability in fuzzy reasoning. The combinations of generalized fuzzy computation, the MAPI model and specific distributed architecture proposed in the next sections, are used as a powerful neurosymbolic processing tool.

Definition 3.1. *The artificial neuron MAPI (Matching, Aggregation, Projection, Inverse-matching neuron) was proposed by da Rocha (1992) and developed under bio-chemical evidences of neuro-physiology:*

$$\text{MAPI} = \{\{X_p\}, Y, T, R, C, \Theta, \{\alpha, f\}\}\,, \quad \text{where} \qquad (3.1)$$

- $\{X_p\}$ is the family of pre-synaptic inputs conveyed over MAPI by all its n pre-synaptic axons;
- Y is the output code of MAPI;
- T is the family of transmitters used by MAPI to exchange messages with other neurons;
- R is the family of receptors to bind the transmitter released by the pre-synaptic neurons; The strength w_i of the synapsis with the ith pre-synaptic neuron is:

$$w_i = M(t)^{\wedge} M(r)^* \mu(t, r)^{\underline{a}} v_0 \qquad (3.2)$$

and the post-synaptic activity v_i is evaluated as:

$$v_i = x_i{}^{\circ} w_i \qquad (3.3)$$

- $M(t)$ is the size of functional pool of the transmitter t at the pre-synaptic cell MAPI_i. $M(r)$ is the amount of r available to bind t. $\mu(t, r)$ is the affinity of $t^{\wedge}r$ binding. v_0 is the standard electromagnetic variation triggered by this binding. $^{\wedge}, ^*, ^{\underline{a}}, ^{\circ}$ are T-norms or T-co-norms;

- Θ is the function used to aggregate the actual pre-synaptic activity;
- $\{\alpha, f\}$ is a family of thresholds and encoding functions defined as:

$$y = \begin{cases} wl, & \text{if} \quad a_{\text{MAPI}} < \alpha 1 \\ wu, & \text{if} \quad a_{\text{MAPI}} \geq \alpha 2 \\ f(a_{\text{MAPI}}), & \text{otherwise} \end{cases} \tag{3.4}$$

- C is the set of controllers, activated by $t_i {}^{\wedge} r_i \gg c_i$, $t_i \in T_p$, $r_i \in R$, $c_i \in C$. T_p is the set of presynaptic transmitters. Each c_i exercises actions over MAPI itself and over other neurons.

The formal neuron introduced through *Definition 3.1* exhibits the capabilities of a multipurpose processing device, since it is able to handle different types of numerical calculations. This is in contrast with and includes the simple processing capability of the classic neuron introduced by McCulloch and Pitts in 1943. The structures presented in the next sections use particularized forms of the neural model in order to perform fuzzy computing by choosing various forms of the parameters given above.

While a standard neural network employs multiplication, addition, and sigmoid transfer function; the hybrid neural network HNN is a neural network with crisp signals and weights, and crisp transfer function (Buckley and Hayashi 1995), which uses:

- T-norms to combine the inputs (usually membership degrees of a fuzzy concept) and weights over the unit interval,
- T-conorms to aggregate the results,
- any continuous function from input to output as a transfer function (all inputs, outputs and weights of the hybrid neural network are real numbers from the unit interval [0,1]

Consequently, a fuzzy neural network FNN (Buckley and Hayashi 1995; Fuller 1999) is a neural network with fuzzy signals and/or fuzzy weights, sigmoid transfer function. All the operations are defined by Zadeh's extension principle (Zadeh 1983).

Definition 3.2. *The standard neuron used in this paper is defined as a particular MAPI unit processing real normalized crisp inputs, weights and outputs for which $w_1 = \alpha_1 = 0$, $w_u = \alpha_2 = 1$, and f is any sigmoid function, while $\circ$ is the product operator, and Θ is the addition operator.*

It is worth to emphasize that, in the neural modules proposed below, all the inputs, outputs and weights are real numbers taken from the unit interval [0,1]. As defined in (Fuller 1999), the processing element of hybrid neural networks is called fuzzy neuron.

Definition 3.3. *The fuzzy neuron (as used in this book) is defined as the generalized MAPI unit, which process crisp and/or fuzzy inputs, weights and* *outputs, using T-norms to combine the inputs and weights over the unit interval, T-conorms to aggregate the results, and any continuous function as a transfer function.*

In this context, the AND fuzzy neuron is a MAPI unit with $w_1 = \alpha_1 = 0$, $w_u = \alpha_2 = 1$, f is the identity function, $\circ$ is T-conorm, and Θ is a T-norm, and produces the output:

$$y_{AND} = \operatorname*{T\text{-}norm}_{i=1}^{p}[\text{T-conorm}(x_i, w_i)] \qquad (3.5)$$

for p inputs. Similarly, the OR fuzzy neuron viewed as a MAPI unit with $w_1 = \alpha_1 = 0$, $w_u = \alpha_2 = 1$, f the identity function, combines the inputs x_i and the weights w_i by a T-norm to produce $a_{OR} = \text{T-norm}(x_i, w_i)$. The input of OR_{MAPI} neuron is aggregated by a T-conorm to produce:

$$y_{OR} = \operatorname*{T\text{-}conorm}_{i=1}^{p}[\text{T-norm}(x_i, w_i)] \qquad (3.6)$$

Thus, if T-norm = min and T-conorm = max, then the OR neuron realizes the max-min composition:

$$y_{OR} = \operatorname*{max}_{i=1}^{p}[(\min(x_i, w_i)] \qquad (3.7)$$

If $w_i = 1$ for all $i = 1, 2, \ldots, p$, then $y_{OR} = \operatorname*{max}_{i=1}^{p} x_i$. Similarly, the MIN and MAX neural operators are defined in (da Rocha 1992). The role of the connection weights is to define the impact of the individual inputs on the output, which is viewed as a result of aggregation. For example, the higher the value w_i, the stronger the impact of x_i on the output of the OR neuron. This result is used to combine the outputs of different implicit and explicit knowledge modules, using a competitive unsupervised-trained gateway network.

Finally, the Implication-OR fuzzy neuron is defined (Fuller 1999) as an unit for which the signals x_i and weights w_i are combined by a fuzzy implication operator Imp to produce $a_i = \text{Imp}(x_i, w_i)$, $i = 1, 2, \ldots, p$. The input information a_i is aggregated by a triangular conorm to produce the output:

$$y_{ImpOR} = \operatorname*{T\text{-}conorm}_{i=1}^{p}[\text{Imp}(x_i, w_i)] \qquad (3.8)$$

The formal neuron exhibits the capabilities of a multipurpose processing device, since it is able to handle different types of numerical calculations. This is in contrast with, and includes, the simple processing capability of the classic neuron introduced by McCulloch and Pitts in 1943. The structures presented in the next sections use particularized forms of the neural model in order to perform fuzzy computing through various forms of the parameters described above.

A standard neural network employs multiplication, addition, and sigmoid transfer function. However, the hybrid neural network HNN is a neural structure with crisp signals and weights; a crisp transfer function, using T-norms to combine the inputs (usually membership degrees of a fuzzy concept), and weights over the unit interval; T-conorms to aggregate the results, and any continuous function from input to output as a transfer function.

3.2 Knowledge Representation in Hybrid Intelligent Systems

The reason for developing hybrid intelligent systems and, more specifically, hybrid neural systems, is based on successful applications of subsymbolic knowledge-based systems, especially Artificial Neural Networks and Fuzzy Neural Networks (Fuller 1999), and on the advantages of symbolic Knowledge-Based Systems (KBS). In Hybrid Intelligent Systems, connectionist tools can be interpreted as hardware and fuzzy logic as software implementation of human reasoning: modular structures encoding explicit and implicit knowledge build homogenous inductive and deductive learning and reasoning systems.

From the point of view of cognitive science, the purely neural representation offers the advantage of homogeneity, the possibility of performing distributed tasks, and working with incomplete and noisy data. From the point of view of KBS, symbolic representations have the clear advantages of human interpretation, explicit control, and knowledge abstraction.

While symbolic systems have a long tradition in Artificial Intelligence research, in the last fifteen years a tremendous amount of research on connectionist fields emerged, partially due to the interesting results in mathematical pattern recognition, to the emergence of software agent technology and its inherent needs to rapidly learn from the available data. In connectionist systems, learning plays a central role, useful when data are abundant and little knowledge is available. This is demonstrated by the good results of high-level connectionism applied to natural language processing as well as commonsense reasoning (Sun 1994), and multi-agent systems (Khosla and Dillon 1997).

In this section, hybrid intelligent systems combining connectionist and symbolic features are presented. Symbolic information is inserted into neural networks, designed to represent domain knowledge, then training examples are used to refine the initial knowledge. Finally, the trained networks process the output for new instances and apply specific methods (Omlin and Giles 1996; Benitez et al. 1997; Jagielska et al. 1997, 1999) to extract symbolic information from the trained neural networks. The knowledge extracted during a final step can explain how the outputs are computed, therefore helping to interpret the refined connectionist knowledge. The connectionist integration of explicit knowledge and learning by examples appears to be the natural solution to developing connectionist intelligent systems.

Explicit and implicit rules are represented in a neural manner using fuzzy neural networks FNN (Fuller 1999), hybrid neural networks HNN (Buckley and Hayashi 1995), multilayer perceptron MLP, or neuro-fuzzy nets, such as those implemented in NEFCLASS (Nauck and Kruse 1998), MPNN (da Rocha 1992), NEIKES (Neagu et al. 2002). While fuzzy logic provides the inference mechanism under cognitive uncertainty, neural nets offer in the connectionist implementation of HIS the advantages of learning, adaptation, fault-tolerance, parallelism and generalization.

3.2.1 The Implicit Knowledge Representation

Definition 3.4. *Implicit knowledge is the knowledge represented by connectionist structures, which are created and adapted by a learning algorithm. The intrinsic representation of implicit knowledge is based on the numerical weights of the connections between neurons. Assuring to integrate implicit knowledge and explicit given fuzzy rules into the same global network, the trained neural networks are chosen as hybrid/fuzzy neural networks.*

We will consider for the sake of generality an Implicit Knowledge Module (IKM) implemented as a multilayered neural structure based on an input layer establishing the inputs to perform the membership degrees of the current values, a fully connected three-layered FNN of type 2 (Fuller 1999), and a defuzzification layer (Fig. 3.3). The FNN of type 2 (FNN2) implements IF-THEN fuzzy rules and is characterized by fuzzy inputs, outputs and crisp

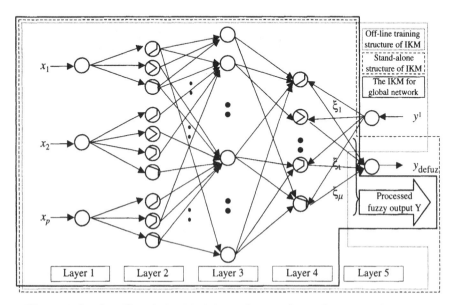

Fig. 3.3. Implicit Knowledge Module implemented as a fuzzy neural network

weights. It is proposed that the input nodes of the FNN2 will be MAPI neurons, parameterized to implement given membership functions of the term
set of each linguistic input. As a generalized approach, in (Jang and Sun
1993; Lin and Lee 1991) the authors proposed similar structures in which the
objective is to approximate the shape of membership functions. The objective of the FNN2 viewed as an IKM is just to learn the fuzzy rules and to
implement the dependences between the linguistic output and the fuzzified
inputs of the system.

The implicit knowledge module is designed with respect to the fact that
the inputs and the output of the system are described as linguistic variables
by the human expert: the specific term sets, their cardinalities, and correspondent membership functions are already established. The characteristics
of the nodes in each layer and the objectives of the layers are:

Layer 1: The nodes in this layer transmit the crisp input values to the
next layer. They are single input standard neurons, with identity activation
function. The weights between this layer and the next one are set to unity.

Layer 2: The nodes in this layer process the membership degrees of the
current input with respect to the $m_i, i = 1, 2, \ldots, n$, a priori described membership functions of the n fuzzy sets defining the fuzzy shape of each input
considered as a linguistic term. In fact, this layer is the fuzzifier part of the
module.

The weights of the connections between layer 1 and layer 2 are set to one.
The linguistic variable X_i is described by m_i fuzzy sets, A_{ij}, having the degrees of membership performed by the functions $\mu_{ij}(x_i)$, $j = 1, 2, \ldots, m_i, i =
1, 2, \ldots, p$. According to the definitions 1 and 3, we can now define the mem-
bership neuron.

Definition 3.5. *The membership neuron is a MAPI (fuzzy) single-input
single-output neuronal unit with $w_1 = \alpha_1 = 0$, $w_u = \alpha_2 = 1$, and f is described by the membership function $\mu_j(x)$, $j = 1, 2, \ldots, m$, where m is the
number of fuzzy terms describing the linguistic variable X.*

Layer 3: (the rule nodes): The outputs of layer 2, the units of layer 3,
and the inputs to the layer 4 build a standard neural network implemented
as a MLP and used to learn the dependencies between the fuzzified forms
of the inputs and the output of the system. Each node of layer 3 could be
considered as defining (Jagielska 1998; Kasabov 1996) a fuzzy rule between
one input and the output. The weights are adapted by a backpropagation of
error algorithm. Once the learning objectives have been achieved, using the
Relative Rule Strength (RRS) method, the Effect Measure Method (Jagielska
1998), or the more general Causal Index Method (Enbutsu et al. 1991), it is
possible to extract some interesting forms of dependencies between the fuzzy
inputs and output of the system (see also Chap. 6 for applications).

Layer 4: Has a double transmission behavior. In the right to the left
transmission, the units are processing, similarly to the layer 2 nodes, the

membership degrees of the current output in order to adapt the weights to the layer 3 with respect of the backpropagation method. In the case of left-to-right transmission, the units perform the fuzzy OR operation in order to integrate the fired rules with the same consequent (as fuzzy OR neurons). The links between layer 4 and layer five are set to one.

Layer 5: Contains two nodes, each one of a different type. The first type is similar to the units from layer one, performing the transmission of the crisp current output value to feed into the MLP-like network. The second type is acting as a defuzzifier. The defuzzification of the outputs follows the methods proposed by da Rocha (1992) using a final controlling neuron. The method of defuzzification determines the programmable profile of the final neuron: center of gravity is implemented by a MAPI averaging device, acting as a numeric neural processor which performs powered averaging functions. Procedures based on max and min values can be implemented with MAX/MIN MAPI neurons (Fig. 3.4), as described in the following equation (defuzzification calculus is performed by the MAPI defuzzifier):

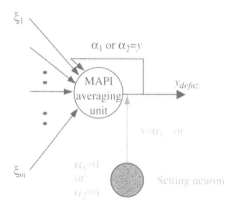

Fig. 3.4. Defuzzifying the output using an averaging device and AND/OR setting neuron in a MAPI implementation (where ξ_i, $i = 1, 2, \ldots, m$, are the computed outputs of the layer 4)

$$y_{\text{defuz}} = \frac{\sum_{\xi \in [0,1]} \xi \cdot \overset{m}{\underset{i=1}{\max}} (\min(\xi_i, \mu_i(\xi)))}{\sum_{\xi \in [0,1]} \overset{m}{\underset{i=1}{\max}} (\min(\xi_i, \mu_i(\xi)))} \qquad (3.9)$$

where y_{defuz} is the crisp normalized value of the output Y, described by m fuzzy sets, B_j, having the degrees of membership performed by the functions $\mu_j(\xi)$, $j = 1, 2, \ldots, m$ (ξ is incrementally describing the interval [0,1]), and ξ_i are the processed membership degrees of the current output.

The described architecture of the IKM (Fig. 3.3) has a structure of a FNN2, both, during the training process (as a five-layered structure with the final transmission neuron on layer 5 – the off-line training structure of IKM) and when it transmits the processed membership degrees to the next levels of the neurosymbolic system (in this case when it uses just the first four layers: the IKM structure to be included in the global network). Alternatively, the five-layered architecture has a HNN structure (as defined in Buckley and Hayashi 1995; Fuller 1999) using the MAPI deffuzifier to define the final layer (the stand-alone structure of IKM), and proposed to compute the crisp normalized value of the output.

3.2.2 The Explicit Knowledge Representation

We define the *explicit knowledge* as a knowledge base represented by neural networks, which are computationally identical to a fuzzy rules set, and are created by mapping the given fuzzy rules into hybrid neural networks. The fuzzy rule set is described as a Discrete Fuzzy Rule-Based System DFRBS (Buckley and Hayashi 1995). The intrinsic representation of explicit knowledge is based on fuzzy neurons in a MAPI implementation. The numerical weights corresponding to the connections between neurons are computed using the Combine Rules First Method (da Rocha 1992; Buckley and Hayashi 1995), or the Fire Each Rule Method (Buckley and Hayashi 1995; Fuller 1999). This approach will assure the maximum compatibility in definition, implementation, integration and processing (including further developments) of various modules of HIS.

Combine Rules First Method
for Explicit Knowledge Representation

The capabilities of MAPI-based HNN to perform fuzzy computing (da Rocha 1991, 1992) are used to implement DFRBS (Neagu and Bumbaru 1999). The neural reasoning engine is accorded to multiple premises fuzzy rules using fuzzy connectives. According to the extended version of Modus Ponens (Zadeh 1983):

$$\frac{\text{IF } X_1 \text{ is } A_1{}^\wedge \ \ldots \ ^\wedge X_p \text{ is } A_p \text{ then } Y \text{ is } B}{(X_1 \text{ is } A_1') \ ^\wedge \ \ldots \ ^\wedge \ (X_p \text{ is } A_p')} \tag{3.10}$$
$$Y \text{ is } B'$$

where X_i, $i = 1, 2, \ldots, p$, are the inputs of the system, and Y is the output of the system, all of them expressed as linguistic variables. The standard implementation of fuzzy sets connectives (Pedrycz 1993) involves triangular norms or co-norms, implemented by the fuzzy neurons, as described above:

$$y = \mathop{\text{T-conorm}}_{i=1}^{p}[x_i \text{T-norm } w_i] \qquad (3.11)$$

$$y = \mathop{\text{T-norm}}_{i=1}^{p}[x_i \text{ T-conorm } w_i] \qquad (3.12)$$

where x_i and w_i are the inputs and the weights of the MAPI neuron implementing fuzzy operators. The chosen neural implementation is an equivalent structure, which uses the method of combining rules first (Buckley and Hayashi 1995).

Let be considered a single rule with two antecedents described as:

$$\text{IF} X_1 \text{ is } A_1 \text{ AND } X_2 \text{ is } A_2 \text{ THEN } Y \text{ is } B \qquad (3.13)$$

where A_1, A_2, B are fuzzy sets having associated matching functions μ_{A1}, μ_{A2}, μ_B. Let the matching function $\mu_{A1}(\xi)$ be described by a vector X_1 of size m_1, so that:

$$x_{1i} = \mu_{A1}(\xi), \text{ if } \alpha_i < \xi \le \alpha_{i+1}, \ i = 1, 2, \dots, m_1 - 1 \qquad (3.14)$$

Thus, the fuzzy set A_1 is (Fig. 3.5):

$$A_1 = [x_{11} \dots x_{1m1}] \qquad (3.15)$$

Similarly, fuzzy sets A_2 and B are described in discrete forms as follows:

$$A_2 = [x_{21} \dots x_{2m2}], \ x_{2i} = \mu_{A2}(\psi), \text{ if } \beta_i < \psi \le \beta_{i+1}, \ i = 1, 2, \dots, m_2 - 1 \qquad (3.16)$$
$$B = [y_1 \dots y_m], \text{ with: } y_i = \mu_B(v), \text{ if } \gamma_i < v \le \gamma_{i+1}, \ i = 1, 2, \dots, m - 1 \qquad (3.17)$$

The fuzzy relation:

$$R : A_1 \times A_2 \times B \to [0, 1] \qquad (3.18)$$

having:

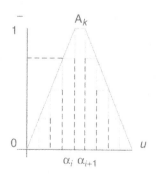

Fig. 3.5. Discrete form of fuzzy set A_k

$$\mu_R(x_1, x_2, y) = (\mu_{A1}(\xi)^\wedge \mu_{A2}(\psi))\Gamma \mu_B(v) \qquad (3.19)$$

defines the implication according to (3.14), where $^\wedge$ is a conjunctive T-norm and Γ is an associative T-norm; so that given A'_1, A'_2:

$$B' = (A'_1 \wedge A'_2)^\circ R \qquad (3.20)$$

where $^\circ$ is usually interpreted as max-Γ operator.

The implementation of an explicit multi-premise rule, given in (3.14), into an equivalent HNN structure using MAPI neurons with fuzzy abilities is shown in Fig. 3.6. The vectors are of size m_1, m_2, m, respectively. The HNN in Fig. 3.6 is equivalent to the DFRBS described by relation (3.21) if: input neurons are used to convert the current values of entries X_1 and X_2 to correspondent values $\mu_{A1}(\xi)$, $\mu_{A2}(\psi)$; the weights between input and associative neurons are set to 1; associative neurons H_{ij} process $(x'_{1i} \wedge x'_{2j})$ if the encoding function is the $^\wedge$ T-norm; the weights between associative and output neurons are: $w_{ijk} = (\mu_{A1}(x_{1i}) \wedge \mu_{A2}(x_{2j}))\Gamma \mu_B(y_k)$; the setting neurons S_i, $i = 0, 1, \ldots, m_1 * m_2$, provide the synchronism $H_{11} < H_{m1,1} < H_{1,m2} < H_{m1,m2}$ "$<$" means *fires before*). In these conditions, $B' = (A'_1 \wedge A'_2)^\circ R$ (da Rocha 1992).

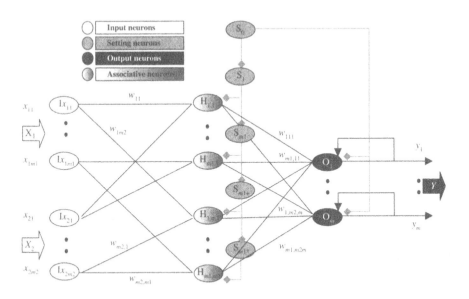

Fig. 3.6. MAPI-based HNN equivalent with a rule with two premises (Combine Rules First Method)

The generalization for a fuzzy rule with n antecedents is implemented in the Multi-Purpose Neural Network (MPNN) structure (da Rocha 1992; Neagu and Palade 1999; Neagu and Bumbaru 1999) by adjusting the correspondent number of input neurons, using a number of associative neurons verifying the constraint: $N_H = m_1{}^* m_2{}^* \ldots {}^* m_n$, and defining the adequate wiring between neurons in these two layers, and between neurons in associative and output layers.

The defuzzification of the output follows the methods proposed by da Rocha (1992), adding to the proposed structure a final controlling neuron. The method of defuzzification determines the programmable profile of the final defuzzifier neuron where the center of gravity is implemented by the MAPI averaging device. Procedures based on max and min values can be implemented with MAX/MIN MAPI neurons, as described in (da Rocha 1992). The structure proposed in Fig. 3.6, completed with a final deffuzifier MAPI-based unit, builds the stand-alone structure of EKM (in fact a structure equivalent to a fuzzy rule-based system).

Fire Rules First Method for Explicit Knowledge Representation

Let's suppose that the fuzzy expert system to be mapped into EKM is described, for simplicity, by one block of rules of the form:

$$\text{IF } X \text{ is } A_{ij} \text{ THEN } Y \text{ is } B_{ik}, i = 1, 2, \ldots, n, \ j = 1, 2, \ldots, m_x, \ k = 1, 2, \ldots, m \tag{3.21}$$

The input to the system is: X is A' with the conclusion: Y is B'. The fuzzy relation R_i modeling the implication in each rule (Buckley and Hayashi 1995) is defined as a function of $A_{ij}(x)$ and $B_{ik}(y)$ for the normalized crisp values x and y in $[0,1]$. For example, this function can be Mamdani's min (Mamdani 1997): $R_i(x, y) = \min(A_{ij}(x), B_{ik}(y))$. Then we compose A' with each R_i to get B'_{ik}, the conclusion of the ith rule according to the compositional rule of inference:

$$B'_{ik} = A' \,{}^\wedge R_i . \tag{3.22}$$

Then the conclusion B' computed by EKM is obtained combining all the local outputs B'_{ik} for all y in $[0, 1]$ using some aggregation operator Γ:

$$B' = \mathop{\Gamma}_{i=1}^{n} (B'i) \tag{3.23}$$

The developed structure is an HNN (Fig. 3.7): all the weights are equal to one, the transfer functions of the neurons in the hidden layer are R_i, the inputs B'_{ik} to the final neuron are aggregated by Γ, while the transfer function of the output neuron is the identity function. The entire structure, including the final neuron, implements the block of rules described in (3.21), while each rule R_i, $i = 1, 2, \ldots, n$, is implemented by the modules EKM$_i$.

For a multi-input system, the described structure and the (3.21), (3.22), (3.23) should be generalized, according to (Buckley and Hayashi 1995). A hybrid neural network computationally identical to a DFRBS with just two discrete inputs is presented in Fig. 3.8. The defuzzification of the output requires a final controlling MAPI-based neuron to the proposed structure, which is adapted to the specific structure as described in (da Rocha 1992), in order to build an EKM stand-alone network.

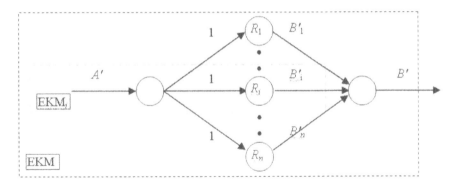

Fig. 3.7. Fire Rules First methods for Explicit Knowledge Representation

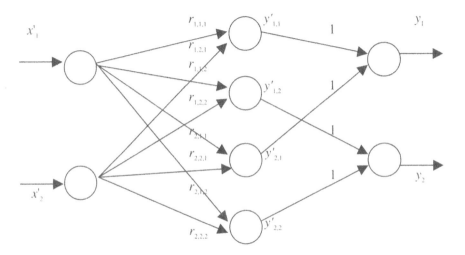

Fig. 3.8. Explicit Knowledge Module EKM computationally identical to DFRBS

3.3 Concluding Remarks

In this chapter, a homogeneous methodology to build HISs, based on a generalized fuzzy neuronal model proposed by da Rocha in 1992 has being highlighted. The concepts of implicit knowledge and explicit knowledge were introduced, taking into account the main principles in Machine Learning modeling, that actual models were introduced as structural mathematical models influenced and simplifying natural entities. Therefore, Hybrid Intelligent Systems can be developed and implemented by connectionist structures, able to further collaborations and integrations in large modular structures.

4 Fuzzy Rules Extraction
from Connectionist Structures

In the conjugate effort of building shells for Hybrid Intelligent Systems with a homogenous architecture, based on neural networks, a difficult task is to exhibit and explain the results of neural calculus as a parallel inference process. This chapter focuses on a strictly fuzzy approach to neural networks. It presents some interactive fuzzy operators in order to extract connectionist-represented knowledge, based on the concept of f-duality. The methodologies are tested on simple and traditional case studies, using two known benchmarks: the Iris problem and the Portfolio problem.

4.1 Introduction

The combinations of generalized fuzzy computation and specific neural architecture are used as a fuzzy-connectionist processing tool for knowledge representation and qualitative reasoning. In the last decade, Artificial Neural Networks (ANN) are not just extensively used as universal approximators (Buckley et al. 1993; Jang and Sun 1993), but, in conjunction with fuzzy logic, are proposed to be, as special architectures (Benitez et al. 1996,1997; Buckley and Hayashi 1995; Langari and Wang 1996), equivalent with Fuzzy Rule Based Systems (FRBS). Some studies (Benitez et al. 1997; Buckley and Hayashi 1995; Neagu et al. 1999; Neagu and Palade 1999; Neagu and Bumbaru 1999) explicitly developed techniques used to describe an ANN as an FRBS. This effort was made in order to give a satisfactory explanation (Kasabov 1996) of neural network behavior as an efficient computing model for solving hard tasks in Artificial Intelligence.

Symbolic processing is considered a traditional way in Artificial Intelligence; unlike symbolic models, learning plays a central role in connectionist structures. A combination of both approaches is already the subject of research in hybrid systems (Hilario 1997; Palade 1999). Viewing connectionist models as a powerful tool to process knowledge, the following step is to try to build connectionist intelligent systems, mainly applied to perceptual tasks, where discovering explicit rules does not seem either natural, or direct. The objective is the uniformity of integration: explicit and implicit rules should be represented in a neural manner.

Mircea Gh. Negoita, Daniel Neagu, and Vasile Palade: *Computational Intelligence: Engineering of Hybrid Systems*, StudFuzz **174**, 41–57 (2005)
www.springerlink.com
© Springer-Verlag Berlin Heidelberg 2005

Architectures combining cooperating connectionist modules (Hashem 1997; Langari and Wang 1996) are proposed to solve integration of explicit and implicit knowledge. Since the discrete fuzzy inputs and outputs are considered common for all rules, a general strategy has been identified (Neagu et al. 1999) that combines: explicit knowledge modules EKM (developed in a top-down manner, using the methods of mapping available explicit rules in neural structures, Neagu and Palade 1999) and implicit knowledge modules IKM (responsible for unmanageable cases of implicit knowledge, achieved using learning by example paradigm). In such a way, modular connectionist architectures can be used to solve in a homogenous manner specific tasks, combining the advantage of learning by example paradigm and FRBS transparency.

In the structures described in this section (see the appendix for details), a MLP approach of implicit knowledge module is proposed to be used in order to improve the behavior of the implicit knowledge modules, as an alternative way for extracting fuzzy rules (Benitez et al. 1997; Neagu and Palade 2000a) from training data sets, and to compare the overall performance of the hybrid system with common approaches. Following the constraints of the methods proposed in (Benitez et al. 1996), the used MLP structure is a multi-input single-output (MISO) network trained off-line, using standard neurons and crisp inputs, weights, and outputs.

4.2 Fuzzy Rules Extraction Using Fuzzy Interactive Connectives

4.2.1 The Notion of f-duality

The studies focused on the equivalence between ANN and fuzzy rule-based systems (FRBS) establish most of the results through an approximation process, with the main disadvantage of exponential increase of needed number of rules or number of required neurons (Benitez et al. 1997; Jagielska et al. 1997; Jagielska 1998; Kasabov 1996). Based on the theoretical results from (Benitez et al. 1996, 1997; Neagu and Palade 2000a), it is possible to build FRBS calculating the same function as an implicit knowledge representation using ANNs. In this case, the concept of f-duality, applied on the three layered feedforward (with/without biases) neural network trained to represent a set of implicit data values, supports the theoretical background to develop a new class of fuzzy connectives (Benitez et al. 1997). The three-layered feedforward network uses a logistic activation function in hidden neurons and identity in output neurons. In the context, the concept of f-duality is introduced (see *Proposition 4.1* and *Definition 4.1* below) in order to define fuzzy operators that enable us to give an explicit interpretation of ANN.

Based on the Theorem 1 (Benitez et al. 1997), f-duality and the methodology presented in the same paper, the idea is to develop specific interactive fuzzy operators, adapted for different kinds of logistic activation function of hidden neurons. The significance of the term *"interactive"* is given by the inputs correlation embedded in the behavior of the fuzzy operator, proven in *Lemma 4.2* and argued in (Benitez et al. 1997). The necessity to describe such kinds of versions of fuzzy operators is to give the possibility to extract fuzzy rules from different versions of trained neural networks.

The artificial neural networks are described by a three-layer feedforward net with a logistic activation function in hidden neurons and identity in output neurons. In the context, the concept of f-duality is introduced by Benitez et al. (1997) in *Proposition 4.1* and *Definition 4.1* in order to define fuzzy operators that enable us to give an explicit interpretation of ANN.

Proposition 4.1. *Let $f : X \to Y$ be a bijective function and let $\oplus$ be an operation defined in the domain of f, X. Then there is one and only one operation, $\otimes$, defined in the range of f, Y, which verifies the relation:*

$$f(x_1 \oplus x_2) = f(x_1) \otimes f(x_2) \tag{4.1}$$

Definition 4.1. *Let f be a bijective function and let $\oplus$ be an operation defined in the domain of f. The operation $\otimes$ whose existence is introduced in the preceding proposition is called the f-dual of $\oplus$.*

4.2.2 Fuzzy Interactive i_{atan}-OR Operator Used for Rules Extraction from ANNs

Let us consider the operation $+$ in $\mathbb{R}$ and the sigmoid function atan: $f_{atansig}(x) = (1/\pi)\mathrm{atan}(x) + 0.5$ (Fig. 4.1), continuous (and bijective) application from $\mathbb{R}$ to $(0, 1)$.

Lemma 4.1. *The $f_{atansig}$-dual of $+$ is $\bullet$, defined as:*

$$a \bullet b = \frac{1}{\pi}\left(\frac{\pi}{2} + \mathrm{atan}\frac{\sin(\pi(a+b-1))}{\cos(\pi(a-0.5))\cos(\pi(b-0.5))}\right), \tag{4.2}$$

Proof. Let a, $b \in (0,1)$ and x_1, $x_2 \in \mathbb{R}$ such that $a = f_{atansig}(x_1)$ and $b = f_{atansig}(x_2)$.

Hence:

$$x = \tan(\pi \cdot f_{atansig}(x) - \pi/2)$$
$$x_1 = \tan(\pi \cdot a - \pi/2)$$
$$x_2 = \tan(\pi \cdot b - \pi/2) .$$
$$x_1 + x_2 = \tan(\pi \cdot a - \pi/2) + \tan(\pi \cdot b - \pi/2)$$
$$= \frac{\sin(\pi \cdot a + \pi \cdot b - \pi)}{\cos(\pi \cdot a - \pi/2) \cdot \cos(\pi \cdot b - \pi/2)}$$

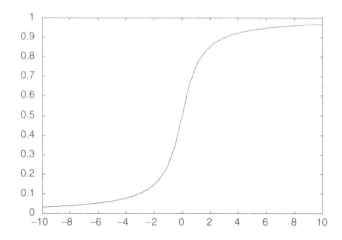

Fig. 4.1. Sigmoid activation function f_{atansig}

By definition of f_{atansig}-dual of the + operator:
$\quad$ a $\bullet$ b $= f_{\text{atansig}}(x_1) \bullet f_{\text{atansig}}(x_2) = f_{\text{atansig}}(x_1 + x_2)$.
$\quad x_1 + x_2 = \tan(\pi \cdot f_{\text{atansig}}(x_1 + x_2) - \pi/2)$.

$$f_{\text{atansig}}(x_1 + x_2) = \frac{1}{\pi}\left(\frac{\pi}{2} + \text{atan}\frac{\sin(\pi(a + b - 1))}{\cos(\pi(a - 0.5))\cos(\pi(b - 0.5))}\right),$$

$$\text{which leads to: } a \bullet b = \frac{1}{\pi}\left(\frac{\pi}{2} + \text{atan}\frac{\sin(\pi(a + b - 1))}{\cos(\pi(a - 0.5))\cos(\pi(b - 0.5))}\right).$$

Definition 4.2. *The operator defined in* Lemma 4.1 *is called* interactive$_{\text{atan}}$-OR, *shortly,* i_{atan}-*OR (Fig. 4.2).*

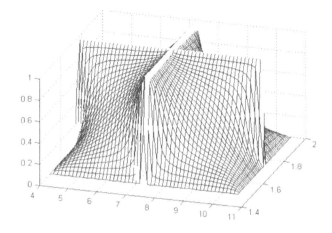

Fig. 4.2. The operator *interactive*$_{\text{atan}}$-OR (in a two inputs application context)

Lemma 4.2. *Let f_{atansig}-dual of $+$ operator, $\bullet$. Then $\bullet$ verifies:*

- is commutative.
- is associative.

the neutral element of $\bullet$ is $e = 1/2$.
  Existence of inverse element: $\forall a \in (0,1), \exists! a' \in (0,1)$ such that $a \bullet a' = e, a' = 1 - a$.

Corollary 4.1. *Let $\bullet$ be f_{atansig}-dual of $+$. Then $((0,1), \bullet)$ is an abelian group.*

Proof. Is known that $(\mathbb{R}, +)$ is an abelian group. Hence, by *Proposition 4.1*, $((0,1), \bullet)$ is an abelian group too: f_{atansig} becomes an isomorphism between the two groups. These conclusions validate *Lemma 4.2* and its *Corollary 4.1*. The neutral element for $\bullet$ is $f_{\text{atansig}}(0) = 1/2$. The inverse element of $a \in (0,1)$ is obtained using the property, established by *Lemma 4.5*: $f_{\text{atansig}}(-x) = 1 - f_{\text{atansig}}(x)$.

Lemma 4.3. *f_{atansig}-dual of $+$ extends to n arguments:*

$$a_1 \bullet a_2 \bullet \cdots \bullet a_n = \frac{1}{\pi} \left(\frac{\pi}{2} + \text{atan}(\tan(\pi(a_1 - 0.5)) + \tan(\pi(a_2 - 0.5)) \right.$$

$$\left. + \cdots + \tan(\pi(an - 0.5))) \right) \tag{4.3}$$

Proof. Trivial using associativity of $\bullet$.

Lemma 4.4. *f_{atansig}-dual of $+$, $\bullet$, verifies relations:*

$$\lim_{\text{a}i \to 0} a_1 \bullet a_2 \bullet \cdots \bullet a_n = 0, \forall a1, \cdots, a_n \in (0,1), \forall i \in \{1, \ldots, n\}$$
$$\lim_{\text{a}i \to 1} a_1 \bullet a_2 \bullet \cdots \bullet a_n = 1, \forall a1, \ldots, a_n \in (0,1), \forall i \in \{1, \ldots, n\}$$

- is strictly increasing in every argument.

Proof. When $a_i \to 0$, then $\tan(\pi(a_i - 0.5)) = -\infty$. Thus:

$$\text{atan} \left(\tan(\pi(a_i - 0.5)) + \sum_{\substack{j=1 \\ j \neq i}}^{n} \tan(\pi(a_j - 0.5)) \right) \to -\pi/2 \,.$$

and

$$a_1 \bullet a_2 \bullet \cdots \bullet a_n \to 0.$$

When $a_i \to 1$, then $\tan \pi(a_i - 0.5)) = \infty$, and:

$$\text{atan}\left(\tan(\pi(a_i - 0.5)) + \sum_{\substack{j=1 \\ j\neq i}}^{n} \tan(\pi(a_j - 0.5)) \right) \to \pi/2 .$$

Thus: $a_1 \bullet a_2 \bullet \cdots \bullet a_n \to 1$.

Let us consider $a_2, \ldots, a_n$ with fixed values in order to prove the assertion for a_1. The expression $a_1 \bullet a_2 \bullet \cdots \bullet a_n$ reduces to $1/\pi(\pi/2 + \text{atan}(\tan(\pi(a_1 - 0.5)) + k)$, with k being a positive constant. Considering $a, a' \in (0,1), a > a'$, is obtained that $\tan(\pi(a_1 - 0.5)) > \tan(\pi(a' - 0.5))$, which implies: $\text{atan}(\tan(\pi(a - 0.5)) + k) > \text{atan}(\tan(\pi(a' - 0.5)) + k)$. Finally: $a \bullet a_2 \bullet \cdots \bullet a_n < a' \bullet a_2 \bullet \cdots \bullet a_n$. By commutativity, the result is extended on the entire relation.

Lemma 4.5. *Let be the sigmoid function* $f(x) = \text{atan}(x)/\pi + 1/2$. *Then* $f(-x) = 1 - f(x)$.

Proof. Function $\text{atan}(x)$ is antisymmetric:

$$f(-x) = -\text{atan}(x)/\pi + 1/2 = 1 - (\text{atan}(x)/\pi + 1/2)$$

Thus: $f(-x) = 1 - f(x)$.

Proposition 4.2. *Let* $(A, *)$, $(B, +)$, $(C, \bullet)$ *abelian groups with the property that exist bijective functions* $f_1 : B \to A, f_2 : B \to C$ *such that:*

$\forall \, x, x' \in B, f_1(x + x') = f_1(x) * f_1(x')$ and $f_2(x + x') = f_2(x) \bullet f_2(x') (f_1, f_2$ are isomorphisms).

Then $\exists! \, f_3 : A \to C$ a bijective function, $f_3 = f_2 \circ f_1^{-1}$ such that: $\forall y, y' \in A, f_3(y * y') = f_3(y) \bullet f_3(y')$ (in other words, the isomorphism from $(A, *)$ to $(C, \bullet)$ is a composition of isomorphisms through $(B, +)$).

Proof. f_1 is a bijective application. By considering *Proposition 4.1*:

$$\forall y, y' \in A, \exists! x, x' \in B \colon f_1^{-1}(y * y') = f_1^{-1}(y) + f_1^{-1}(y') .$$

Let $x = f_1^{-1}(y)$, $x' = f_1^{-1}(y')$.

$$\forall x, x' \in B, \exists! z, z' \in C \colon f_2(x + x') = f_2(x) \bullet f_2(x') .$$

Let $z = f_1^{-1}(x)$, $z' = f_1^{-1}(x')$.

These lead to:

$$\forall y, y' \in A, f_2(f_1^{-1}(y) + f_1^{-1}(y')) = f_2(f_1^{-1}(y)) \bullet f_2(f_1^{-1}(y')) .$$

Then: $f_1^{-1}(y * y') = f_1^{-1}(y) + f_1^{-1}(y')$.

Hence:

$$(A4) f_2(f_1^{-1}(y * y')) = f_2(f_1^{-1}(y)) \bullet f_2(f_1^{-1}(y')),$$

where $f_2(f_1^{-1}(y))$ is a bijective application from A to C, composition of bijective functions.

Lemma 4.6. f_{atansig}-*dual of* $+$, $\bullet$, *is functional equivalent with* f_A-*dual of* $+$, $*$, *through logistic function* $f_A(x) = \frac{1}{1+e^{-x}}$ *(see for details Benitez et al. 1997).*

Proof. Trivial, by *Definitions 4.1, 4.2, Lemma 4.5* and the definition of i-OR operator (Benitez et al. 1997).

Fuzzy Rules Extraction from ANN

Based on the properties of i_{atan}-OR, *Lemma 4.5*, *Lemma 4.6*, and the theorem proving the equivalence between a trained feedforward neural network with biases and a fuzzy additive system described in (Benitez et al. 1996, 1997) by:

$$R_{jk}: \text{IF } \sum_{i=1}^{n} x_i w_{ij} + \tau_j \text{ is } A_{jk} \text{ THEN } z_k = \beta_{jk} \tag{4.4}$$

the fuzzy rule "x_i is A_{jk}^i" must be interpreted as:

"x_i is greater than approximately $r/w_{ij} - \tau_j$" (if $w_{ij} > 0$), or:
"x_i is lower than approximately $-(r/w_{ij} - \tau_j)$" (if $w_{ij} < 0$),

where r is a positive real number obtained from a α-cut (for example, 0.9).

Since the concept of f-duality is general, it could be used to produce other interactive operators: i-OR (Benitez et al. 1997), i_{tanh}-OR (Neagu and Bumbaru 1999, Fig. 4.3), or some conjunctive forms such as i-AND (Fig. 4.4). The main applications of these operators are knowledge acquisition refinement and explanation of neural inferences.

4.2.3 Fuzzy Interactive i_{tanh}-OR Operator Used for Rules Extraction from ANNs

In the context of *Proposition 4.1* and *Definition 4.1*, let us consider the operation $+$ in $\mathbb{R}$ and the sigmoidal function tansig: $f_{\text{tansig}}(x) = \frac{2}{1+e^{-2x}} - 1$ (Fig. 4.5), continuous (and bijective) application from $\mathbb{R}$ to $(-1, 1)$.

Lemma 4.7. *The* f_{tansig}-*dual of* $+$ *is* $\circ$, *defined as:*

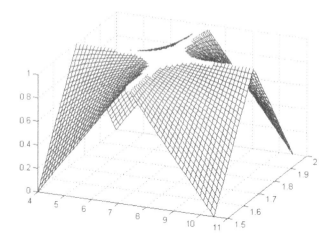

Fig. 4.3. *Interactive*$_{\text{tanh}}$-OR operatot (in the Iris application context)

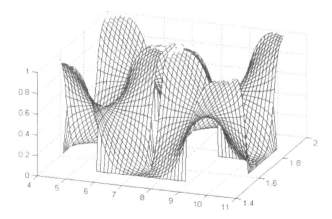

Fig. 4.4. *Interactive* AND operatot (in the Iris application context)

$$a \circ b = \frac{a + b}{1 + ab} \, . \tag{4.5}$$

Proof. Let $a, b \in (-1, 1)$ and $x_1, x_2 \in \mathbb{R}$ such that $a = f_{\text{tansig}}(x_1)$ and $b = f_{\text{tansig}}(x_2)$.

Hence:

$$x = -\frac{1}{2} \ln \left(\frac{1 - f_{\text{tansig}}(x)}{1 + f_{\text{tansig}}(x)} \right), \text{ so:}$$

$$x_1 = -\frac{1}{2} \ln \left(\frac{1 - a}{1 + a} \right), \quad x_2 = -\frac{1}{2} \ln \left(\frac{1 - b}{1 + b} \right) \, .$$

$$x_1 + x_2 = -\frac{1}{2} \ln \left(\frac{1 - a}{1 + a} \cdot \frac{1 - b}{1 + b} \right) \, .$$

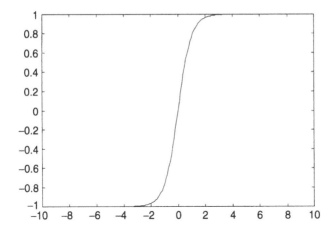

Fig. 4.5. Sigmoid activation function f_{tansing}

By definition of f_{tansig}-dual of the $+$ operator:

$$a \circ b = f_{\text{tansig}}(x_1) \circ f_{\text{tansig}}(x_2) = f_{\text{tansig}}(x_1 + x_2) \ .$$

$$x_1 + x_2 = -\frac{1}{2} \ln \left(\frac{1 - f_{\text{tansig}}(x_1 + x_2)}{1 + f_{\text{tansig}}(x_1 + x_2)} \right) x_1 + x_2$$

$$= -\frac{1}{2} \ln \left(\frac{2}{1 + f_{\text{tansig}}(x_1 + x_2)} - 1 \right) \ .$$

$$\frac{1-a}{1+a} \cdot \frac{1-b}{1+b} = \frac{2}{1 + f_{\text{tansig}}(x_1 + x_2)} - 1 \ ,$$

which leads to: $f_{\text{tansig}}(x_1 + x_2) = \frac{a+b}{1+ab}$.

Definition 4.3. *The operator defined in* Lemma 4.7 *is called interactive*$_{\text{tanh}}$-*OR, shortly,* i_{tanh}-*OR (Fig. 4.5).*

Lemma 4.8. *Let* f_{tansig}-*dual of* $+$ *operator,* $\circ$. *Then* $\circ$ *verifies:*

1. $\circ$ is commutative.
2. $\circ$ is associative.
3. the neutral element of $\circ$ is $e = 0$.
4. Existence of inverse element: $\forall a \in (-1,1), \exists! a' \in (-1,1)$ such that a $\circ a' = e, a' = -a$.

Corollary 4.2. *Let* $\circ$ *be* f_{tansig}-*dual of* $+$. *Then* $((-1,1), \circ)$ *is an abelian group.*

Proof. Is known that $(\mathbb{R}, +)$ is an abelian group. By *Proposition 4.1*, $((-1,1), \circ)$ is an abelian group too: f_{tansig} becomes an isomorphism between the two groups. These conclusions validate *Lemma 4.8* and its *Corollary 4.2*. The neutral element for $\circ$ is $f_{\text{tansig}}(0) = 0$. The inverse element of $a \in (-1,1)$

is obtained using the property established by *Lemma 4.10*: $f_{\text{tansig}}(-x) = -f_{\text{tansig}}(x)$.

Corollary 4.3. *Let* $\circ$ *be* f_{tansig}-*dual of* $+$. *Then* $((0,1), \circ)$ *is an abelian semigroup.*

Proof. $\circ$ is a compositional rule on $(0,1)$ because:

for $\forall a, b \in (0,1), a + b - 1 - ab = (a-1)(1-b) < 0$, so $\dfrac{a+b}{1+ab} < 1$,

(obviously, positive values).

Associativity is derived from: $\dfrac{\dfrac{a+b}{1+ab} + c}{1 + \dfrac{a+b}{1+ab}c} = \dfrac{a + \dfrac{b+c}{1+bc}}{1 + a\dfrac{b+c}{1+bc}}$.

Proving the commutativity is similar.

Lemma 4.9. f_{tansig}-*dual of* $+$ *extends to n arguments:*

$$a_1 \circ a_2 \circ \cdots \circ a_n = \dfrac{\displaystyle\sum_{\substack{j=1 \\ j\,\text{impar}}}^{\leq n} \left(\displaystyle\sum_{\substack{ik=1 \\ i_1 \neq \cdots \neq ij}}^{n} a_{i1} \cdots a_{ij} \right)}{1 + \displaystyle\sum_{\substack{j=2 \\ j\,\text{par}}}^{\leq n} \left(\displaystyle\sum_{\substack{ik=1 \\ i_1 \neq \cdots \neq ij}}^{n} a_{i1} \cdots a_{ij} \right)} \qquad (4.6)$$

Proof. Trivial using associativity of $\circ$.

Lemma 4.10. *Let consider the sigmoid function* $f(x) = 2/(1 + e^{-xw}) - 1$. *Then:*

$$f(-x) = -f(x) . \qquad (4.7)$$

Proof. $f(x) = \dfrac{1 - e^{-xw}}{1 + e^{-xw}}$.

Thus: $f(-x) = \dfrac{1 - e^{xw}}{1 + e^{xw}} = \dfrac{e^{xw}(e^{-xw} - 1)}{e^{xw}(e^{-xw} + 1)} = -\dfrac{1 - e^{-xw}}{1 + e^{-xw}}$.

Therefore: $f(-x) = -f(x)$.

Note that the mapping operation of the function f_{tansig} in $(0,1)$ have consequently as a result the formula of the sigmoid function:

$$f(x) = (1 + f_{\text{tansig}}(x))/2 = 1/(1 + e^{-2x}) = f_A(2x) ,$$

where f_A is the logistic function: $f_A(x) = \dfrac{1}{1 + e^{-x}}$ (Benitez et al. 1996).

Fuzzy Rules Extraction from ANNs

Based on the properties enounced in *Lemma 4.9, Lemma 4.10, Corrolary 4.2* and *Corrolary 4.3* and the theorem proving the equivalence between a trained feedforward neural network with biases and a fuzzy additive system described in (Benitez et al. 1996, 1997) by (4.4), the fuzzy rule "x_i is A^i_{jk}" must be interpreted as:

"x_i is greater than at least $r/w_{ij} - \tau_j$" (if $w_{ij} > 0$), or:
"x_i is lower than at most $-(r/w_{ij} - \tau_j)$" (if $w_{ij} < 0$),

where r is a positive real number obtained from a α-cut (for example, $\alpha = 0.9, r = 1.47$).

4.3 Case Studies to Extract Fuzzy Rules Using Fuzzy Interactive Connectives

The two interactive operators are tested in two well known case studies: the Iris problem and the Portfolio problem.

4.3.1 Interactive Operators for the IRIS Problem

The goal of well-known iris problem is to recognize the type of an iris plant to which a given instance belongs. The data set is composed of 150 records, equally distributed between three classes: setosa, versicolor, and virginica. Two classes are not linearly separable from each other, while the third is linearly separable from the others. The data set is characterized by four attributes: petal length, petal width, sepal length, and sepal width, hence ANN is using four input neurons.

i_{atan}-OR Operator for the IRIS Problem

The three possible classes are coded as values in $(0, 1)$: 0.1, 0.5 and 0.9 respectively, such that the application required a single output neuron. The activation function of hidden neurons is sigmoid atansig function f_{atansig}. Obviously, the number of obtained rules is equal to number of hidden neurons. Hence, for three fuzzy rules, we trained a feedforward network with four input neurons, three hidden neurons, and one output neuron (Fig. 4.6).

The input/hidden weights matrix after supervised learning is:

$$W^T = [w_{ij}]^T = \begin{bmatrix} 1.8465 & 5.5406 & 2.7217 & 2.8102 \\ 2.3933 & 2.5549 & -1.6371 & 2.1335 \\ 2.1572 & -6.0267 & 3.2005 & -1.1336 \end{bmatrix}.$$

The hidden/output weights matrix is:

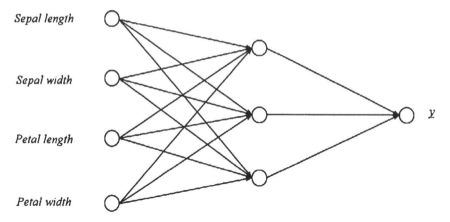

Sepal length

Sepal width

Petal length

Petal width

y

Fig. 4.6. The ANN for the IRIS Problem

$B^T = [\beta_{jk}]^T = \begin{bmatrix} -2.6372 & -0.7893 & -4.7428 \end{bmatrix}$, while the biases for hidden neurons are: $T^T = [\tau_{jk}]^T = \begin{bmatrix} -3.6773 & 0.0046 & 3.4464 \end{bmatrix}$.

The extracted rule set equivalent to trained ANN, obtained on the basis of the above-anounced mechanism, is:

R1.
IF *sepal-length* is greater than approximately 2.160
i atan-OR
sepal-width is greater than approximately 0.720
i atan-OR
petal-length is greater than approximately 1.465
i atan-OR
petal-width is greater than approximately 1.419
THEN y=-3.6773.
R2.
IF *sepal-length* is greater than approximately 1.282
i atan-OR
sepal-width is grater than approximately 1.201
i atan-OR
petal-length is not greater than approximately 1.874
i atan-OR
petal-width is greater than approximately 1.438
THEN y=0.0046.
R3.
IF *sepal-length* is greater than approximately 1.023
i atan-OR
sepal-width is not greater than approximately 0.366
i atan-OR

petal-length is greater than approximately 0.690
i atan-OR
petal-width is not greater than approximately 1.948
THEN *y*=3.4464.

The process of classification for a given input is determined by an aggregation computation using the interactive operator. The instance is matched against the rule premises, each rule being fired to a certain degree v_j. The global output is the weighted sum of these degrees: $y = -3.6773v_1 + 0.00462v_2 + 3.4464v_3$. The class chosen for a given instance is that with the closest numerical value to y.

i_{tanh}-OR Operator for the IRIS Problem

In the same context defined by Fig. 4.6, the input/hidden weights matrix after supervised learning is:

$$W^T = [w_{ij}]^T = \begin{bmatrix} 0.1149 & 0.6656 & 0.9809 & -1.4081 \\ -0.8745 & -1.2795 & 0.5231 & 0.8482 \\ 0.0995 & 1.3551 & -0.1533 & -1.2349 \end{bmatrix}$$

The hidden/output weights matrix is:

$$B^T = [\beta_{jk}]^T = \begin{bmatrix} -0.4334 & -0.9620 & -0.8430 \end{bmatrix},$$

while the biases for hidden neurons are:

$$T^T = [\tau_{jk}]^T = \begin{bmatrix} -1.8549 & 0.0001 & 1.8425 \end{bmatrix}.$$

The extracted rule set equivalent to the trained ANN is:

R1.
IF *S*epal length is greater than at least 16.829
i tanh-OR
 Sepal width is greater than at least 2.905
i tanh-OR
Petal length is greater than at lest 1.971
i tanh-OR
Petal width is lower than at most -1.373
THEN *y*=-0.4334.
R2.
IF *S*epal length is lower than at most -1.680
i tanh-OR
Sepal width is lower than at most -1.14887
i tanh-OR
Petal length is greater than at least 2.810
i tanh-OR

```
Petal width is greater than at least 1.733
THEN y=-0.962.
R3.
IF Sepal length is greater than at least 10.144
```
i_{tanh}-OR
```
Sepal width is greater than at least 0.744
```
i_{tanh}-OR
```
Petal length is lower than at most -6.584
```
i_{tanh}-OR
```
1••imea petalei is lower than at most -0.817
THEN y=-0.843.
```

The process of classification for a given input is determined by an aggregation computation using the interactive operator. The instance is matched against the rule premises, each rule being fired to a certain degree v_j. The global output is the weighted sum of these degrees: $y = -0.4334v_1 - 0.962v_2 - 0.843v_3$. The class chosen for a given instance is that with the closest numerical value to y.

4.3.2 Interactive Operators for the Portfolio Problem

The goal of the portfolio problem (Fuller 1999) is to predict the portfolio value taking into account the currency fluctuations on the global finance market. The data set is composed of 141 records, equally distributed across financial year 1997. The data set is characterized by three currency rates (USD/DEM, USD/ECU and USD/1000ROL).

The ANN designed to learn the portfolio data set is a feedforward three-layered neural network 3-2-1 (three input neurons, two hidden neurons and one output neuron for portfolio value, as a combination of currencies deposit, Fig. 4.7). The activation function of the hidden neurons is f_{atansig}.

The input/hidden weights matrix after supervised learning is:

$$W^T = [w_{ij}]^T = \begin{bmatrix} -2.8799 & 5.5696 & 1.4336 \\ -1.2003 & -0.2842 & 2.0518 \end{bmatrix}.$$

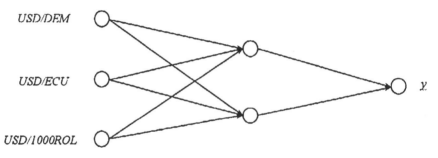

Fig. 4.7. The ANN for the Portfolio Problem

The hidden/output weights matrix is:

$$B^T = [\beta_{jk}]^T = \begin{bmatrix} -60.2425 & -50.4879 \end{bmatrix},$$

while the biases for hidden neurons are:

$$T^T = [\tau_{jk}]^T = \begin{bmatrix} 3.7397 & -1.9506 \end{bmatrix}.$$

The extracted rule set equivalent to the trained ANN, obtained on the basis of the above-enounced mechanism, is:

R1. IF *USD/DEM* is greater than at least 1.127
    i_{tanh}-OR
    USD/ECU is lower than at most -2.695
    i_{tanh}-OR
    USD/1000ROL is lower than at most -3.371
    THEN $y_2 = -4.3181$.
R2. IF *USD/DEM* is lower than at most -1.497
    i_{tanh}-OR
    USD/ECU is greater than at least 0.782
    i_{tanh}-OR
    USD/1000ROL is lower than at most -1.352
    THEN $y_2 = -3.2509$.

The process of classification for a given input is determined by an aggregation computation using the interactive operator.

The instance is matched against the rule premises, each rule being fired to a certain degree v_j. The global output is the weighted sum of these degrees:

$$y = -4.3181v_1 - 3.2509v_2 .$$

i_{atan}-OR Operator for the Portfolio Problem

The input/hidden weights matrix after supervised learning is:

$$W^T = [w_{ij}]^T = \begin{bmatrix} 2.7545 & -1.5095 & -1.4781 \\ -3.0961 & 1.6610 & -0.3162 \end{bmatrix}.$$

The hidden/output weights matrix is:
$B^T = [\beta_{jk}]^T = \begin{bmatrix} -3.8360 & 4.1859 \end{bmatrix}$, while the biases for hidden neurons are:

$$T^T = [\tau_{jk}]^T = \begin{bmatrix} -3.5425 & -3.5277 \end{bmatrix}.$$

The extracted rule set equivalent to the trained ANN, obtained on the basis of the above-announced mechanism, is:

R1.
IF *USD/DEM* is greater than approximately 1.543
i_{atan}-OR
USD/ECU is not greater than approximately 2.816
i_{atan}-OR
USD/1000ROL is not greater than approximately 2.876
THEN y_2=-3.8360.
R2.
IF *USD/DEM* is not greater than approximately 1.371
i_{atan}-OR
USD/ECU is greater than approximately 2.556
i_{atan}-OR
USD/1000ROL is not greater than approximately -13.428
THEN y_2 = 4.1859.

The process of classification for a given input is determined by an aggregation computation using the interactive operator. The instance is matched against the rule premises, each rule being fired to a certain degree v_j. The global output is the weighted sum of these degrees: $y = -3.8360v_1 + 4.1859v_2$.

i_{tanh}-OR Operator for the Portfolio Problem

The input/hidden weights matrix after supervised learning is:

$$W^T = [w_{ij}]^T = \begin{bmatrix} -2.8799 & 5.5696 & 1.4336 \\ -1.2003 & -0.2842 & 2.0518 \end{bmatrix} .$$

The hidden/output weights matrix is:

$$B^T = [\beta_{jk}]^T = \begin{bmatrix} -60.2425 & -50.4879 \end{bmatrix} ,$$

while the biases for hidden neurons are:

$$T^T = [\tau_{jk}]^T = \begin{bmatrix} 3.7397 & -1.9506 \end{bmatrix} .$$

The extracted rule set equivalent to the trained ANN, obtained on the basis of the above-announced mechanism, is:

R1.
IF *USD/DEM* is greater than at least 1.127
i_{tanh}-OR
USD/ECU is lower than at most -2.695
i_{tanh}-OR
USD/1000ROL is lower than at most -3.371
THEN y_2 = -4.3181.
R2.

IF *USD/DEM* is lower than at most -1.497
i_{tanh}-OR
USD/ECU is greater than at least 0.782
i_{tanh}-OR
USD/1000ROL is lower than at most -1.352
THEN y_2 = -3.2509.

The process of classification for a given input is determined by an aggregation computation using the interactive operator. The instance is matched against the rule premises, each rule being fired to a certain degree v_j. The global output is the weighted sum of these degrees: $y = -4.3181v_1 - 3.2509v_2$.

4.4 Concluding Remarks

Based on the equality between ANN and FRBS, and the concept of f-duality introduced by Benitez et al. (1997), this chapter introduced two interactive fuzzy logic operators i_{atan}-OR and i_{tanh}-OR. The interactive operators enable us to reformulate fuzzy rules into a more compact as well as comprehensible way. The knowledge the neural network acquires during the learning process is represented as a fuzzy rule set. This constitutes an interpretation of ANNs: building a particular fuzzy rule based system that acts exactly the same as a neural network.

In this chapter the mathematical relationships between the proposed fuzzy operators and the i-OR operator proposed in (Benitez et al. 1997) was also proved. The result is an easy-to-understand interpretation of the knowledge acquired in a trained neural network. Two well-known benchmarks, the Iris problem and the Portfolio problem, were used to exhibit the usefulness of the proposed fuzzy operators.

5 Integration of Explicit and Implicit Knowledge in Hybrid Intelligent Systems

5.1 Introduction

The introduction of modular networks into fuzzy systems provides new insights into the integration of explicit and implicit knowledge in a connectionist representation. The modular network is a connectionist architecture that allows each module to exhibit its own "opinion" about the entries, in order to classify or predict the output. Thus, a modular network offers several advantages over a single neural network in terms of learning speed, generalization and representation capabilities (Haykin 1994; Kosko 1992; Jacobs et al. 1991).

Hence, the idea to represent explicit and implicit knowledge in a connectionist manner is based on the concept of modularity (Haykin 1994; Jacobs et al. 1991). Modularity may be viewed as a manifestation of the "divide and conquer" principle, which let us to solve complex computational tasks by dividing the problem into simple subtasks and then combining their individual solutions. We use the modular network concept to integrate explicit and implicit knowledge formally defined as follows (a definition adapted from Hashem 1997):

A neural network is said to be modular if the computation performed by the network can be decomposed into two or more modules that operate on inputs of the main problem without communicating with each other. The outputs of the modules are mediated by an integrating unit that is not permitted to feed information back to the modules. In particular, the integrating unit decides how the outputs of the modules should be combined to form the final output of the system.

The main approaches of learning paradigms are involved in a modular network as follows (Jacobs et al. 1991; Hashem 1997): unsupervised learning allows modules to compete with each other to produce the output, while supervised learning uses an external teacher that supplies the desired target patterns to train different modules.

The global network (GN) of our approach (Fig. 5.1) is a modular structure including two different "points of view" about the same problem: the implicit knowledge, implemented by the trained neuro-fuzzy network, and the explicit knowledge, represented by a collection of special neural networks

Mircea Gh. Negoita, Daniel Neagu, and Vasile Palade: *Computational Intelligence: Engineering of Hybrid Systems*, StudFuzz **174**, 59–69 (2005)
www.springerlink.com
© Springer-Verlag Berlin Heidelberg 2005

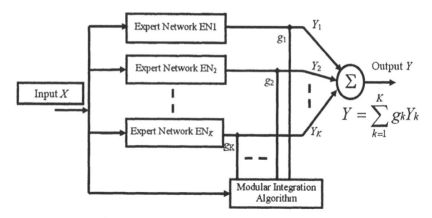

Fig. 5.1. The architecture of an integrated modular implicit and explicit knowledge-based system

equivalent to some rules proposed by human experts (Neagu and Palade 1999). The first type of modules (IKM) is responsible for generalization and processing noisy cases, using implicit knowledge, achieved through learning from examples. The second one (EKM) is developed in a top-down manner, using the methods of mapping available explicit rules in hybrid neural structures, as described above.

Architectures based on cooperating connectionist modules are proposed to solve integration of explicit and implicit knowledge. In this section, we propose to review three strategies for combining IKM and EKM in order to build a global hybrid system: the Fire Each Module (FEM), the Unsupervised-trained Gating Network (UGN), and the Supervised-trained Gating Network (SGN). The first strategy is an adapted Fire Each Rule method (Buckley and Hayashi 1995) in the context of modular networks. The second strategy proposed a competitive-based aggregation of the EKM and IKM outputs, while the third strategy uses a supervised trained layer to process the overall output of the modules.

5.2 A Formal Description of Implicit and Explicit Knowledge-Based Intelligent Systems

The modular HIS considered in this chapter is a multi-input single-output neuro-fuzzy system (MISO). Its general goal is to model a combination of data and expert information to relate some inputs with the corresponding output value:

$$\Phi : D \subseteq \boldsymbol{R}^n \to \boldsymbol{R} , \tag{5.1}$$

where $n \in N$ is the number of the inputs for the application domain.

This leads to the following steps in a fuzzy neural computational process: (a) development of individual knowledge-based connectionist models, (b) modeling synaptic connections of individual models, to incorporate fuzziness into modules, (c) adjusting (Neagu and Palade 2002) the ensemble voting algorithm (Fig. 5.1). In Fig. 5.1, the output layer (formally described by a sum sign) depicts any algorithm to combine expert modules "opinions" about the particular problem and situation they were trained for, including any weighted voting, average or other statistical combination, fuzzy inference or supervised/unsupervised trained gating connectionis approach.

Let's consider a MISO HIS with n inputs. Let also consider $U = \prod_{i=1}^{n+1} D_i$ the universe of discourse over the application domain as the Cartesian product of sets $D_i, i = 1 \ldots n + 1$, for the input variables $X_i \in D_i, i = 1 \ldots n$, and the output $Y \in D_{n+1}$. A HIS integrated model of the problem Φ, based on implicit (IKM) and explicit knowledge (EKM) modules, is a good approximation of Φ if:

$$HIS = \left\{ M_j \underset{j=1\ldots m}{} / \forall \varepsilon > 0, \exists X \in \prod_{i=1}^{n} D_i, \forall Y = \Phi(X) : \|M_j(X) - Y\| < \varepsilon \right\}$$

(5.2)

where the knowledge modules are functional models:

$$M_j : \prod_{i=1}^{n} D_{ij} \to D_{n+1,j}$$

(5.3)

The modules M_j are, in our approach, either implicit or explicit knowledge models:

$M_j \in \{MIKM\_CNN, MIKM\_FNN, MEKM\_Mamdani, MEKM\_Sugeno\}$. For any of these M_j models, we can propose, following (5.1)–(5.3), a formal parameter-based description of modular HIS:

$$M_j = \langle \Theta, \Lambda, \Omega \rangle$$

(5.4)

where Θ is the set of *topological parameters* (i.e. number of layers, number of neurons on each layer, connection matrices, type and number of individual models and gating networks), Λ is the set of *learning parameters* (learning rate, momentum term, any early stopping attribute for implicit knowledge modules, but NIL for explicit knowledge modules) and Ω is the set of *description parameters* (type of fuzzy sets, parameters of membership functions associated to linguistic variables).

Three distinctive cases to develop integrated HIS models are identified:

Case 1: $D_j = \prod_{i=1}^{n} D_{ij}$ for all $j = 1 \ldots m$: a modular architecture (Neagu and Palade 2002) of expert on the whole input domain.

Case 2: $I_{j=1}^{m} \prod_{i=1}^{n} D_{ij} = 0$ and $D_i \cap D_j = 0$, for $j, k = 1, \ldots, m$. The HIS model is a collection of m expert models on disjunctive input domains; the system is a top-down integrated decomposition model, by dividing the initial problem in separate less-complex sub-problems (Neagu 2002).

Case 3: $I_{j=1}^{m} \prod_{i=1}^{n} D_{ij} \neq 0$: models built on overlapping sub-domains; further algorithms to refine the problem as cases 1 or 2 are required (Neagu et al. 1999).

5.3 Techniques of Integrating the Implicit and Explicit Knowledge into Hybrid Intelligent Systems

So far, few strategies to combine IKM and EKM in a global HIS have been proposed (Neagu and Palade 2002): Fire Each Module (FEM), Unsupervised-trained Gating Network (UGN), Supervised-trained Gating Network (SGN), majority voting etc. FEM is an adapted Fire Each Rule method (Buckley and Hayashi 1995) for modular networks, in two versions: statistical combination of crisp outputs (FEMS) or fuzzy inference of linguistic outputs (FEMF). UGN proposes competitive aggregation of EKMs and IKMs, while SGN uses a supervised trained layer to recognize component outputs.

5.3.1 Fire Each Module Method for Integration of Neural Implicit and Explicit Knowledge Modules

The proposed Fire Each Module (FEM) strategy is the simplest mode to integrate IKM and EKM with fuzzy output. The general approach of this modular structure is proposed in (Neagu et al. 2002a) and shown in Fig. 5.2. After off-line training phase is applied to the implicit neuro-fuzzy module, the general output of the system is composed as a T-conorm (Pedrycz 1993) of the fuzzy outputs of each module: the four-layered IKM structure for the global network and the EKM (implemented using combine rules first or fire each rule method). The system is viewed as equivalent to a set of given fuzzy

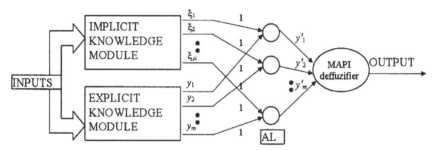

Fig. 5.2. Integration of explicit and implicit knowledge modules in the global network according to the FEM strategy

rules: the overall output is computed using firing (implicit and explicit) rules first method (Buckley and Hayashi 1995). The method of combining the specific membership degrees provided by both the IKM structure (ξ_i values, $i = 1, 2, \ldots, m$) and the EKM structure (y_i values, $i = 1, 2, \ldots, m$), would be done component wise so let:

$$y'_i = \text{T-conorm}(\xi_i, y_i), i = 1, 2, \ldots, m \qquad (5.5)$$

for some aggregating operator, in particular the max fuzzy operator. In the hidden aggregative layer (AL), all the weights are set to one and the neurons aggregate the specific computed membership degrees ξ_i and y_i as implicit. The respective explicit opinion about the current output is described with $B_i^{-\text{th}}$ fuzzy term (where the terms set describing the output is $B = \{B_1, \ldots, B_i, \ldots, B_m\}$). Practically, the inputs for the MAPI deffuzifier describe the shape of the fuzzy output.

The final neuron is a MAPI device, which computes the crisp value of the output, using, for example, the center of gravity method.

The methodology proposed to build the global network architecture is based on obtaining fuzzy rules, describing the system, which are obtained from a human expert or from a set of examples of its behavior. Basically, the methodology consists of:

1. Identification of input and output linguistic variables. The variables are represented by fuzzy sets that are mapped in MAPI units.
2. The IKM described in the Sect. 3.2.1 is built and train as a five-layered fuzzy neural network (the off-line training structure of IKM).
3. From the hidden network of the IKM, we extract the most relevant rules, using the Relative Rule Strength method and either the Effect Measure Method (Jagielska 1998) or the Causal Index Method (Enbutsu et al. 1991).
4. We construct a set of possible explicit rules in a given problem with the help of a human expert, using both, external rules and those already extracted at step 3, as the most voted and trusted dependencies between the inputs and output. We map all these rules into the EKM as described in Sect. 3.2.2. Some explicit rules could have just a part of identified inputs in the rule premise, represented as active neurons, while the rest of the input neurons will be set as inactive.
5. The four-layered IKM and the EKM, already presented structures (without the deffuzifier MAPI final neuron), are embedded into the architecture described in Fig. 5.2 for which the combining hidden layer AL and the deffuzifier MAPI-based unit are adapted.
6. After an incremental loop sequence based on steps 2 to 5 (which could be used as a knowledge acquisition procedure), the global network is ready to be used as a classifier or prediction tool.

The incremental loop sequence consisting of steps 2 to 5 could be refined on the basis of combining the already given fuzzy rules and training data set as follows. IKM is designed by mapping some external fuzzy rules in the hidden HNN, which further learning with training samples is based on. This way the knowledge is kept at the sub-symbolic level. The main goal of the approach is not just to reduce the training period, but also to improve the generalization abilities of the network. The disadvantages consist in both the redistribution of symbolic a priori knowledge (or at least building haloes of initial rules) and necessity of a new refinement of final incorporated knowledge in the resulted network. This strategy follows the variations of *concept support techniques* (Prem et al. 1993; Wermter and Sun 2000) with the difference of the method used to insert a priori knowledge:

1. Inserting some rules describing a subset of cases of desired input-output mapping, and learning the training samples (inserted explicit rules play the role of a complement of the training sets in supplying knowledge to the network).
2. Inserting the symbolic concepts believed to be relevant for the problem solution and training by supporting the relevant concepts.
3. Inserting explicit rules as in (b), followed by a training phase, in which the used hidden units are different from those designed in first phase.

5.3.2 Unsupervised-trained Gating Network Method for Integration of Neural Implicit and Explicit Knowledge Modules

The proposed structure is based on the modular networks paradigm, by considering the basic configuration consisting of two general types of networks: expert networks (implemented by EKM and neuro-fuzzy IKM) and a gating network GN. A classical modular network considers expert networks competing to learn the training patterns and the gating network mediating the competition (Jacobs et al. 1991; Langari 1993; Haykin 1994). The proposed modular architecture uses neural explicit and implicit knowledge modules, and the gating network for voting the best combination of fuzzy terms computed by expert networks, in order to describe the linguistic output (Fig. 5.3).

The EKM and IKM structures are developed and, respectively, trained. The gating network is also trained, with the constrain to have as many output neurons as there are fuzzy terms chosen to describe the linguistic variable Y as the output of global network. The specific membership degrees provided by both the IKM structure (ξ_i values, $i = 1, 2, \ldots, m$), and the EKM structure (y_i values, $i = 1, 2, \ldots, m$), are aggregated according to the (5.5) by MAPI-based neurons implementing MAX T-conorm (aggregation layer AL). The goal of the learning algorithm for the gating network is to model the distribution of the membership degrees computed by the EKM and the IKM.

The gating network consists of a single layer of m output neurons (Hashem 1997), each one having m inputs. The activation function of its output neurons is a *softmax* transformation (Bridle 1990). The process of gating network training considers that, for each vector $[x'_1, \ldots, x'_m]$ processed by AL, the activation g_i of the ith output neuron is related to the weighted sum of the inputs applied to that neuron. Consequently, the activations of the output neurons in gating network are nonnegative and sum to one:

$$0 \le g_i \le 1 \text{ and } \sum_{i=1}^{m} g_i = 1 \tag{5.6}$$

The additional advantage gained by using the gating network is the implicit defuzzification of the overall output of the system:

Proposition 5.1. *Let $[x_1, \ldots, x_p]$ be the current input of the system and $[y'_1, \ldots, y'_m]$ be the current output of the aggregation layer and lets consider the gating network already trained using the unsupervised compet algorithm (Hagan 1996). Then the overall output y of the UGN system, computed by a softmax transformation is a crisp value representing the defuzzified output of the model.*

Proof. Let's consider that the output of the system is computed in respect with the Sugeno model (the consequent part of each rule is described by a linear regression model (Sugeno and Kang 1988; Takagi 1994)):

$$R_i: \text{ IF } X_1 \text{ is } A_{i1} \text{ AND } X_2 \text{ is } A_{i2} \text{ AND } \ldots \text{ AND } X_p \text{ is } A_{ip}$$

$$\text{THEN } y'_i = \sum_{j=1}^{p} b_{ij} x_j \tag{5.7}$$

where $A_{i1}, A_{i2}, \ldots, A_{ip}$ are fuzzy sets having associated matching functions $\mu_{Ai1}, \mu_{Ai2}, \ldots, \mu_{Aip}, b_{ij}$ are real-valued parameters, y'_i is the local output of the model due to rule R_i, $i = 1, 2, \ldots, m$. The total output of the Sugeno model is a crisp value defined by the weighted average:

$$y = \frac{\sum_{i=1}^{m} h_i y'_i}{\sum_{i=1}^{m} h_i} \tag{5.8}$$

The weight h_i implies the overall truth value of the premise of rule R_i for current input, and is calculated as:

$$h_i = (\mu_{Ai1}(x_1)^\wedge \mu_{Ai2}(x_2)^\wedge \ldots^\wedge \mu_{Aip}(x_p)\,, \tag{5.9}$$

where $^\wedge$ is a conjunctive T-norm. The output y described in (5.8) is a crisp value.

Let's now consider the input vector $[x_1, \ldots, x_p]$ applied to the system described in Fig. 5.3. Then, the output of the entire architecture is:

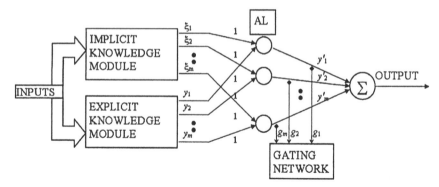

Fig. 5.3. Integration of explicit and implicit modules using an unsupervised-trained gating network (UGN strategy)

$$y = \sum_{i=1}^{m} g_i y'_i . \tag{5.10}$$

The expressions of the current output of the Sugeno model and the proposed structure are similar: each rule in the Sugeno model could be considered an explicit rule into EKM, or a particular way involving one hidden neuron through the IKM, while the relative weight of the ith neuron in the gating network is:

$$g_i = \frac{h_i}{\sum_{i=1}^{m} h_i} . \tag{5.11}$$

In essence, the gating network, proposed to combine the outputs of the aggregating layer, acts as a special defuzzifier.

The methodology proposed to build the global network architecture is partially similar to the FEM methodology, and consists of:

1. Steps 1 to 4 are similar to those described in the previous (Sect. 5.3.1).
2. The four-layered IKM and the EKM, already described structures (without the deffuzifier MAPI final neuron) are embedded into the architecture described in the Fig. 5.3, for which the combining hidden layer AL and the deffuzifier MAPI-based unit are adapted.
3. The gating network is trained (*compet* algorithm) using the AL outputs computed for the training data set of the system.
4. After an incremental loop sequence based on the first step (which could be considered as a knowledge acquisition procedure), the global network is ready to be used as a classifier or prediction tool: the final crisp value of the output is computed using the gating network based on the *softmax* transformation, as described in (5.10).

5.3.3 Supervised-Trained Gating Network Method
for Integration of Neural Implicit
and Explicit Knowledge Modules

The proposed structure contains expert networks represented by a defined number of EKMs and IKMs solving various sub-problems of the main task, and a supervised trained network mediating their outputs' combination. EKMs represent explicit rules, identified by an expert, or refined from a previous knowledge acquisition phase. IKM structures are useful in the overall architecture because of their generalization and processing noisy data abilities.

After training, different expert networks compute different functions, each of them mapping different regions of the input space. Each defuzzified output of the expert networks is considered an input for the final layer. The supervised training process of the final network assures, in fact, a weighted aggregation of expert networks' outputs with respect to their specialization (Fig. 5.4).

The methodology proposed to build the global network architecture using a supervised-trained gating module consists of the following steps:

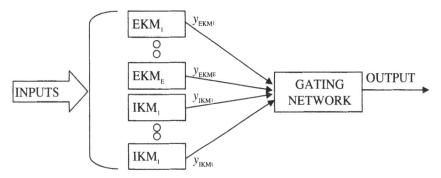

Fig. 5.4. Integration of explicit and implicit modules using a supervised-trained gating network (SGN strategy)

1. Identification of input and output linguistic variables. The variables are represented by fuzzy sets mapped in MAPI units.
2. The IKM modules are represented as HNN and/or MLP networks. We build and train each IKM described in Chap. 3 as a five-layered fuzzy neural network or as a MLP-based structure, in order to assure for each one a crisp specific output.
3. The most relevant rules are extracted from the IKM structures, using the Relative Rule Strength method, the Effect Measure Method or the Causal Index Method in the case of HNN implementation, respectively using interactive fuzzy operators in the case of MLP implementation.

4. We construct a set of possible explicit rules in the given problem with the help of a human expert, using both external rules and those already extracted at step 3 as the most voted and trusty dependencies between inputs and output. We map each rule into a specific EKM structure as described in Sect. 3.2 (using the particularized MAPI-based network depicted in Fig. 3.6 or the EKM$_i$ structure described in Fig. 3.8). A MAPI-based defuzzifier as a final layer completes all the EKM structures in order to assure a crisp output for each module.
5. The IKM and EKM structures are embedded into the global architecture (Fig. 5.4).
6. The gating network is supervised trained using the EKM and IKM computed outputs, so that the overall network is a combination of expert modules.
7. After an incremental loop sequence based on first four steps (which can be considered as a knowledge acquisition procedure), the global network is ready to be used as a classifier or prediction tool. The final crisp value of the output is computed using the gating network which acts as a classifier of the best combination of each expert network's behavior, describing the overall output for a given vector of input data.

5.4 Concluding Remarks on Explicit and Implicit Knowledge Integration

The proposed structures and methods argue the use of connectionist systems in symbolic processing. Since the presented EKMs were demonstrated to be identical to Discrete Fuzzy Rule-based Systems (Bradley et al. 2000; Higuchi et al. 1994) the homogenous integration of explicit rules and training data sets permits better cover of the problem domain.

In that case, the constraint of the size of neural networks is solved by the modularity paradigm. EKMs represent explicit rules identified by an expert or refined from IKM structures. IKMs are useful especially for such complex problems described by (noisy) data sets. The EKM and IKM combination encourages compact solutions for problems described by both data sets distributed in compact domains in the hyperspace, and isolated data, situated in intersection of compact sub-domains or inhomogeneous intervals. After training, different expert networks compute different functions mapping different regions of the input space.

The different sources of the information explicitly and/or implicitly integrated in the presented modules exhibit the problem of knowledge redundancy in the final structure. The proposed methods, based on explicit and implicit module integration, can operate with redundant knowledge spread in both ways of representation. This redundancy should be clearly minimized. The method to implement redundancy minimization is based on selecting specific data sets from the training collection, which are not suitable to verify the implemented rules. The resulting training data sets describe such domains in the hyperspace, which are not covered by the explicit rules. The main disadvantage is that the IKMs are able to generalize just in their domain.

The modular EKM and IKM combination encourages compact solutions for problems described by both data sets distributed in compact domains in the hyperspace and isolated data, situated in intersection of compact subdomains or inhomogeneous intervals. After training, different expert networks compute different functions mapping different regions of the input space. The proposed methods are already applied with promising results in short term air quality prediction (NEIKES: Neagu et al. 2002) and toxicity and carcinogenicity modeling and prediction.

6 Practical Implementation Aspects Regarding Real-World Application of Hybrid Intelligent Systems

This chapter is focussed on the application aspects of *HIS* engineering. The main application areas of *HIS* are mentioned. A lot of outstanding applications could be reviewed, but this is beyond the topic of this book. We tried to introduce just some special original applications. On the other hand, this chapter will also be focused on some combined approaches which involve the main *CI/AI* techniques, but at the same time are both fundamental and with a very large applicability.

Some special application areas are illustrated by two *HIS* as follows:

- *NEIKeS* (**N**eural **E**xplicit and **I**mplicit **K**nowledge Based **S**ystem) – an original Fuzzy Knowledge-Based System for air quality prediction (Neagu et al. 2002);
- *NIKE* – a further NEIKeS application to Predictive Toxicology (Neagu and Gini 2003);
- *WITNeSS* (**W**ellington **I**nstitute of **T**echnology **N**ovel **E**xpert **S**tudent **S**upport) – an intelligent tutoring system (Negoita and Pritchard 2003a,b)

A set of main remarks will be made regarding the combined approaches with large applicability, namely with respect to the optimisation techniques in *HIS*. *GA* can be seen as a better alternative optimisation method to *NN*. *GA* based fusion and transformation systems are increasingly used in most applications. So *GA-FS* reliant *HIS* are used for automating and optimisation tasks in *FS* design. *GA-FS* reliant or even *GA-NN-FS* reliant *HIS* are used for simultaneously optimising the connection weights as well as the *NN* structure and topology. Two applications of *VLGGA* (**V**ariable **L**ength **G**enotype **G**enetic **A**lgorithm) – a special non-standard *GA* – are introduced in the form of two *hybrid intelligent optimisation methods* of large applicability:

- a *VLGGA* based method of learning (from examples) the parameters featuring the fuzzy inference systems (Fagarasan and Negoita 1995);
- a *VLGGA* relied optimisation method of parameters featuring the fuzzy recurrent neural networks (Arotaritei and Negoita 2002)

Mircea Gh. Negoita, Daniel Neagu, and Vasile Palade: *Computational Intelligence: Engineering of Hybrid Systems*, StudFuzz **174**, 71–150 (2005)
www.springerlink.com © Springer-Verlag Berlin Heidelberg 2005

6.1 An Overview on Application Area
of Hybrid Intelligent Systems

Real-world application requests are the engine leading compels the further development of first hybridisation level *HIS* and impose the further developments with respect to *HIS* of second and third level of hybridisation. No area of our social-economic life is developing without the involvement of *HIS* technologies.

A key domain of business activity is the *finance/banking* engineering. *HIS* were efficiently applied in the *area of services* techniques, even to help banks decide whether to offer credit to a customer and assist supermarkets to organize their displays. The business world employs *HIS* techniques in financial engineering both to improve business fitness and to ensure its survival in competitive markets.

The main *HIS* application areas for finance/banking and other business activities are as follows (Goonatilake and Treleaven 1996; Negoita 2003):

- retail banking (mortgage evaluation; forecasting the product demand on the world/national market)
- diverse aspects of marketing activity (customer profiling; cross selling; segmentation tasks – market and customer segmentation; target identification)
- insurance (risk evaluation; premium calculation; aid insurance and warranty claims; automobile insurance and credit card transaction risk assessment)
- investments (asset forecasting; profolio management; the term structure of interest rates and models of investor behavior)
- financial planning with respect to retail outlet location and/or product distribution
- banking security/surveillance (detection of credit application fraud; insider dealing detection; control of credit card use; currency recognition)
- other aspects of financial analysis and prediction (forecasting bankruptcy, credit scoring, securities trading)

On the other hand, a lot of *industrial applications* are the result of *HIS* techniques applied by a few companies throughout their operations. For example:

- generation of *control strategies* for industrial processes (typical to food and chemistry industry, but to other industries too, see optimization of *continuous casting of steel*, for example)
- direct adaptive controllers or hierarchical controllers able to generate a knowledge base from scratch and eliminate the need of a plant model as a consequence
- kinematics and/or sensorial for robot-like systems
- industrial equipment inspection/diagnosis and quality control intelligent systems

- discovery of prediction rules for very complex industrial problems with multiple implications in other area of our social-economic life. An example of such an application was developed in case of a neuro-fuzzy *HIS* that predicted the derived chemical properties of ashes resulted from the combustion processes for thermo-electric generation plants (Castellano et al. 2003); the results having impact on environmental protection and are also supportive for material reuse purposes
- intelligent electric power production and distribution (load forecasting systems, optimization of economic dispatch problem and unit commitment problems, intelligent assistance in loading power overhead transmission lines in different astronomic and meteor-climatic conditions).
- general-purpose *HIS* scheduling systems for air traffic management, maintenance scheduling in power systems, optimization of manufacturing processes/job shop scheduling
- optimisation of delivery systems by using optimised truck routing
- fashion shoe or clothes design systems with an intelligent input to a wider variety of ideas that the designer has access to at any given time, so the response time to market is improved
- fruit storage for the fruit market and food industry, so wilting or fruit diseases are skipped by a suitable storage environment
- complex and accurate diagnosis systems in avionics, electronics equipment
- aerospace application – attitude determination and control of a spacecraft
- sophisticated but intelligent systems for telecommunications solving signal analysis, noise elimination, data compression and skipping possible information flow conflicts on data busses. Fault-free prediction for channel assignment in cellular mobile radio systems may be performed by *HIS*, as well as effective equalization schemes of communication channels in adverse environments, or frequency assignment in satellite communications systems, namely the minimization of the co-channel interference between geo-stationary satellites

A wide range of *HIS* applications are used with respect to speech and handwritten word recognition, including complex talking bilingual dictionaries that recognize spoken word and phrases, then adapts in real time to new accents and dialects.

Typical *HIS* in the environmental sector are performing:

- risk evaluation, resource management
- chemical/pollution evaluations, sonic pollution control and evaluation in metropolis
- weather forecasting (including prediction of zone rainfall or strong storms/ typhoons), waste water treatment plants
- seismic warning and trace editing systems

The important area of public health – medicine – is using *HIS* applications for a large variety of tasks as:

- direct intelligent assisted surgical interventions
- forecast/prediction tasks aimed to support medical diagnosis and prevention (some illustrative examples: reduction of the breast biopsies number needed for evaluating suspected breast cancer; childhood obesity prediction based on medical and sociological data)
- investigation methods reliant on *HIS* image processing (in most of the cases these are strictly very specialized applications. For example: adaptive extraction of nucleus region from stained cancer image to support the computer-aided histological inspection; EEG pattern recognition of epileptiform waves)

But *HIS* applications HAVE evolved so far that They are able to improve the lifes of human beings even in the most unexpected aspects. Take clothing *comfort* for example: wear trials and evaluation using *HIS* led to a precise clothing comfort predictability taking into account subjective factors such as sensation.

In recent years, *HIS* were successfully focused on developing intelligent vehicles, those can provide vehicle information to drivers. At the same time they have developed systems to increase comfort for the motorist. A central role in driver's comfort is played by vehicle health – the current technical condition of the car – based on sensors' outputs regarding the air pressure in the tires, coolant level, engine oil level, break oil level, fuel condition, etc... An *interesting GA–FS* reliant *HIS* provides for a comfortable, successful and relaxed journey as proposed in (Bajaj and Keskar 2003), by maintaining intelligent judgment and control of vehicle health, while at the same calculating the safest distance to be traveled by the car to complete its journey.

Knowledge discovery in databases represents an *HIS* application area of increasing interest to any company, no matter they are involvedis in business or manufacturing (Khosla and Dillon 1997). Different kinds of data are captured in huge data bases as required by the particular profile of activity: a huge amount of information with respect to customers, markets, competitors are of interest to business development companies; information about performances and optimization opportunities as well as any potential source of process improvements or troubleshooting problems is collected for manufacturing purposes. But data collection will be a useless process without the ability to extract knowledge at a later date. All theses aims are covered under the global idea of identifying valid, novel potentially useful and ultimately understandable knowledge extracted from data. The knowledge discovery procedure involves different steps, one of them – data mining namely-being in fact a suitable framework for *HIS*. Technically speaking, *HIS's* role during the data mining step consists of applying data analysis and learning algorithms that practically leads to a particular collection of data patterns (or models).

The most known examples of *HIS* application for knowledge discovery in data-bases are for different difficult complex forecasting tasks. This is the case of financial trading where a continuously learning of new knowledge is required because of "floating" currencies. Also electrical power distribution is reliant on *HIS* technologies for many kinds of different forecasts: a short term forecast as for yearly budget estimation; long term forecasting – with respect to energy, revenue, customer number growth; a forecast for customer energy usage for pricing purchasing, distribution planning, determination of distribution charges.

Another recent but promising area of *HIS* applications is focused on intelligent tutoring systems. Intelligent tutoring systems are a highly effective *HIS* reliant technology for computer teaching systems. They model instructional and teaching strategies, empowering educational programs with the ability to decide on what and how to teach students. A "stand alone" intelligent (*HIS* reliant) tutoring component is added to the usual learning environments, so the work that is done by lecturers and students is complemented (Negoita and Pritchard 2003a,b).

Different applications of *HIS* within the area of advanced Internet technologies solve a large spectrum of refined technical requirements with regard to the quality of services for real-time applications. Actually real-time connectionless technology is largely applied over Internet; see for example in the case of Voice over IP. The quality of service in the case of Voice over IP was solved by using an intelligent hybrid neuro-fuzzy approach: the per-application load balancing is performed with the help of an adaptive feedback control that uses requested quality of service requirements for Voice over IP and optimized link usages over the uncertain patterned Internet traffic (Chimmanee et al. 2003).

6.2 WITNeSS – (Wellington Institute of Technology Novel Expert Student Support) – An Intelligent Tutoring System (ITS)

The human society is evolving, that means the world is changing at a faster and faster rate. These changes are envisaging a drastic change of main real-world social-economic areas such as: business, education, home life, employment, management and a lot of other sectors of activity. But these goals are reliant with no exception on how the latest advances in information technology and computer science can help develop better learning systems. Such a strategy might be implemented by building ultimate *HIS* based learning systems (machines) that are primarily designed to help any category of students become better learners. These artificial learning systems will do nothing else than bringing students into a close encounter with the most important factor in any situations – themselves. There would seem to be no boundaries that would restrict ultimate learning systems, because knowledge and technology will always be advancing – making possible more and more advancements.

Technological support is crucial if the field of electronics is to make hardware architectures as malleable as software. The complexity of the evolving systems is not seen as a problem because the only limits to electronics design are the limits imposed by our own understanding (Negoita and Stoica 2004).

Although in *HIS CI/AI* techniques will be the main field of knowledge involved in development of ultimate learning systems, other disciplines will be embedded too: psychology, human-computer interface technologies, knowledge representation, databases, system analysis and design and last but not least, advanced computer programming technologies.

Usually, learning systems require that the student change to fit the system but with the advent of electronic learning systems, an added flexibility gives the opportunity for *the learning system to change to fit the student's needs.* So ultimate learning systems must interact with a student in a way that is different from the usual. The key role for accomplishing this task inside the ultimate learning systems is played by *HIS* technologies.

The main above mentioned intelligent *"change to fit the students needs"* learning strategy of ultimate learning machines must perform the following tasks in support of students:

- present the student with any content or skill set they wish to learn, in a way that *suits their particular personal, individual learning style and psychological features.* This is a crucial task to be performed by personalization research seeking to develop techniques for learning and exploiting student individual work style preferences and psychological features to deliver *the right content to the right user in the right form at the right time* (Smyth 2003)
- advise the student, when requested by the student, on *how to best learn* the content or skills and help them work out a *suitable study schedule* to handle the content
- the learning system will *co-work with the student in monitoring* the learning schedule
- the monitoring of student's learning schedule must be integrated into a process of *collaborative knowledge* because the students must be aware of others' activities and that the collaboration with other persons (mainly other students or instructors) must be regulated. Different hybrid intelligent systems are to be developed both to support awareness and to regulate the collaboration. This role of monitoring the collaboration is performed by this hybrid intelligent multi-agent technology essentially by computing statistics, detecting possible conflicts (problems) and giving advice synchronously and asynchronously to the students and instructors based on their activities and requests (Chen 2003)
- an *intelligent interactive analysis* is performed on *what the student is doing* and *real-time diagnostic help* is provided when needed. So, when requested by the student, some sources of further help must be provided in real-time, for example: teachers, source materials that the school has, source

material from outside the school, online help, or even other students who
have displayed the knowledge and who have expressed that they will or
might help by request
- an *intelligent content-schedule fitting analysis* must be optionally performed (by any individual student request): the ultimate learning system
  analyses, when asked, how a particular content area fits in with the student's global (total) schedule and the student is accordingly advised how
  his learning might be improved
- the task of monitoring the learning process is a multi-component task.
  One of its sub-task is that a *regular or emergency report* (as the case)
  would be sent electronically to the person directly responsible for tutoring
  the student; this report will completely cover all dynamic aspects of the
  student's learning, for example what help the student asked for, a complete
  log of the students home worked exercises, etc ...

A basic structural block diagram of an ultimate intelligent artificial learning systems would consist of the following units:

- the *Human – Machine Interface* – working out the bi-directional inter-
  communication between different categories of potential users (students,
  teachers, desk support people, even parents) and the component parts of
  the artificial learning system. The human-machine interface must be a hy-
  permedia one, to be connected to different multimedia systems so that
  even the ways are paved for special communications connections (of the
  "teleconference type"). Elements of affective computing would feature in
  this ITS architectural block even by embedding typical design policies into
  five sense communication systems, namely small-scale five sense communication systems that can always be used on a PC. Usually such kinds
  of systems are useful to determine a person's situation in a remote place
  (Yoshino et al.2003), but they can be adapted so as to feature the psychological component of the student model in Web-based educational systems.
  The human-machine interface would also be designed to allow learning by
  mobile phone, so the students can use the mail function of the WWW to
  manage their learning. They would be able to access a suitable common
  schedule, help one another both by registering questions and exchanging
  information of common interest. The interest, motivation and competitive
  behaviour between students may be monitored by a function to display
  relative student progress in a graph form on the WWW (see Yoshida et al.
  2003).

It is envisioned that a large spectrum functionality will be implemented.
The Tutor would have a high profile presence on the WWW so that students
who are looking for a Tutor would have available a virtual classroom providing
a wide range of teaching and learning resources. This would include software
that would empower the Tutor to deal with groups of students, helping them
to work together, providing an intelligent network so that teachers can get

in touch with teachers, parents sharing with other parents, students with students.

- *Artificial Learning Unit* – a unit where any topic to be learned can be processed so that it becames a suitable learning module that can be plugged into the artificial learning system
- *HIS decision making unit* – this is the "brain" of the whole system, having as its main function the intelligent decision making on how to respond to a large variety of user requests
- the *Information Storing Unit* – is a collection of databases with respect to all information aspects involved in the learning process, for example: student profile (features), knowledge pieces of content, general or specific teaching strategies, etc. The role of this collection of databases may turn out to be more important then originally expected. If the databases (teachers, students, parents and content) are properly structured and intelligently connected, then the ultimate learning system's ability to keep people in touch would be greatly enhanced. This will be essential to providing suitable on-line access to the highest quality learning packages and virtual environments for people of all ages and needs for future personal development.

Most of the *I*ntelligent *T*utoring *S*ystems are actually implemented in the form of web-based computing systems for distance learning. This kind of web-based distance education features some natural limitations such as: student's feeling of isolation; the difficulty of a personalised learning addressing the individual needs of each student; the cost of instructors' communication. By a personalised learning that an *ITS* must provide, not only the provision of relevant information to the right student (at the right time and pace) is meant, but also the provision of potential to collaborate with peers is supposed. The enhancement of student's learning experience is achieved in web-based *ITSs* not only by timely recommendations on content and relevant methodologies, but also by facilitating collaboration and communication between similar users of the system. Groups of similar users are identified based on behaviour and rating criteria. Relevant results regarding personalised recommendations, both on course content and on peer-peer groups were cited in (O'Riordan and Griffith 2003) by using an approach that combines information retrieval methods and collaborative filter techniques. Most of these web-based *ITS* were designed to support students, mainly focussed on what will the learning system do for the student. This is also the aim of *WITNeSS*.

But some recent research is also focussed on supporting instructions in how to manage the distance learning courses effectively (Kosba et al. 2003). A Teacher Advisor framework was developed reliant on a fuzzy methodology of extracting student, group and class models. These models are extracted from the tracking information generated by the Web Course Management Systems – a platform commonly adopted in many educational institutions. These models generate the needed advice for course instructors or facilitators, so the teachers in turn provide more effective guidance in diminishing the student's isolation.

The main aim of an *ITS* is to act as a student or teacher advisor, using student modelling techniques to ensure the student's needs are addressed. But modelling the student is characterised by a high degree of uncertainty, so it would seem that the most suitable modelling strategy for achieving any kind of *ITS* assessment or pedagogical function, would to be used. Therefore, the student model used by *WITNeSS* for student diagnosis was built with the understanding that it be a compulsory component of the *ITS*, both for Teacher Advisor and for Student Advisor framework. It also meant that fuzzy modelling techniques be used for carrying out the required functions of an *ITS*.

6.2.1 Introduction on ITS

Education is a real factoring the progress and development of the contemporary society and students play a key role. Students are the main beneficiaries of the educational system and the present and future of human society depends on them. We need to develop in students the ability to grasp new concepts, make better choices and even develop new forms of thinking, learning and problem solving. This would empower them to go on learning and adapting throughout life in a world that is changing at a faster and faster rate. Turning students into creative learners is a fantastic challenge.

The energy reservoir of social development is constantly being fed by the common characteristic featuring of most students: their readiness to learn and their enthusiastic huge appetite to do so. However, the fact that so many students fail who shouldn't, is a clear indication that many have problems and struggle with their learning. Some of these students have great potential and in most ways would be considered good students, but still a great number of them drop out. The explanation of this phenomenon is that they don't have the learning skills needed to succeed. The problem is founded in the intrinsic psychology nature of the human being. They are easy-going, and they come to their tertiary education just expecting everything to be done for them. In other words they come ill-equipped to handle the academic rigors they will face. These students, usually given the problem and guidelines in how to solve it, not only find themselves faced with a new, complexing content to learn, but they also must determine the problem and work out their own guidelines on how to solve it. It is hard to criticize them for their inability to think because they simply haven't been taught how to. They just hadn't been prepared to be independent learners who can think and problem solve.

In the recent years communication and computer technology (including intelligent technologies) has advanced drastically, so highly effective technology has moved out from the research laboratories, resulting in commercial products to be used for real-world educational purposes: **I**ntelligent **T**utoring **S**ystems (*ITS*). Education is increasingly employing *ITS* both for modelling instructional and teaching strategies and for empowering educational programs (Negoita and Pritchard 2003b).

ITS are a highly effective technology that complements the work done by lecturers and tutors by adding a stand-alone intelligent tutoring component to the current learning environments (Negoita and Pritchard 2003a).

6.2.2 Why ITS Are Required

The period of the late 1980s and early 1990s was crucial for learning and education because the experience gained by people acting in this area led to two important conclusions:

- the first conclusion was that students receiving one-to-one instruction performed better than students receiving more traditional teaching (Bloom 1984)
- the second conclusion was that there was no way human teachers could possibly teach every student one-to-one (Redfield and Steuck 1991)

Researchers started to change learning and education methodology; they tried to solve it. In this way *ITS* become a main application area of *AI/CI* techniques and of *HISs*.

By definition *ITSs* are computing teaching systems that model instructional and teaching strategies, empowering educational programmes to decide what and how to teach the students (Wenger 1987; Ohlsson 1987). *ITSs* are of real help in meeting a crucial common challenge that is of interest to all kinds of educational institutions, namely how they can provide effective instruction to students with a diverse range of abilities, interests and needs. Collaboration between the *ITS* software package and the students provides an improved student learning experience by giving the student the ability to not only question the learning content, but to determine how the learning material should be presented to them. This collaborative work dynamically creates a variety of interactive tutorial experiences for any student, because the *ITS* can assess a student's mastery of subject, match this against a knowledge base of learning content and draw from its database of teaching strategies.

In conclusion, research in the area of *ITS* has greatly matured. There is no doubting the important fact that effective communication forms a strong basis for teaching and learning. Therefore a prime educational goal for *AI/CI* in education must be the development of natural language processing. In *AI* technology what was first thought to be easy has turned out to be more difficult and what was through difficult turned out to be easy. For example the algorithmic nature of game playing in large spaces was easily accomplished. Whereas in the 1950s natural language processing was thought to be an easy accomplishment, history has proven otherwise. It is most difficult.

6.2.3 The Basic ITS Structure

A good description of the basic structure of an *ITS* was made in (McTaggart 2001) as illustrated in Fig. 6.1. Here are shown the main typical components to an *ITS*. When these components work together, the framework of a system is created that can recognise patterns of learning behaviour and react to them. The main components of an *ITS*, as described by (McTaggart 2001), are:

1. The Expert Model
2. The Student Model
3. The Instruction Model
4. The Interface Model

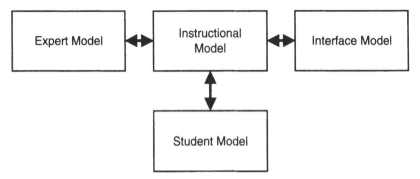

Fig. 6.1. The basic schematic for a typical ITS (Intelligent Tutoring System)

Expert Model. This component organises a database of declarative and procedural knowledge by capturing an expert's knowledge. Expert system techniques and semantic networks are used to achieve this. The *ITS* exposes these knowledge structures to the students. So an opportunity is given to the students both to reflect and then structure their own representation of that knowledge (Orey and Nelson 1993).

Student Model. The main aim of this component is to guide the student through the system's knowledge base. The Student Model stores data about the student's learnt knowledge and also about the student's behaviour as a result of interacting with the *ITS*. The Student Model provides the Instructional Model with the information that enables it to make decision on "what" and "how" to present learning activities to the student. The quality of data held in the Student Model is crucial for the *ITS's* ability to make the above-mentioned decisions and thus provide individualised instructions. This is because data held in the Student Model is constantly being modified as the students interacts with the *ITS*. In this way, the *ITS* will always have a good idea of what the student understands, knows and can do.

Instructional Model. This *ITS* component has the function of intelligent decision making about "what" and "how" to present to the student. The

Instructional Model has this capability because it holds knowledge about tutoring tactics. Ideally, because of the close relationship with the other *ITS's* components, these decisions will always be based on an ever adjusting student profile as from (Buiu 1999).

Interface Model. The human-computer interface is the border across which communication and information flow, implementing a student's window to the "mind" of *ITS*. The main tasks of this interface are: to anticipate the user's actions, to be consistent, to provide a high level of interaction and make use metaphor, see (Orey and Nelson 1993); to be a communication medium, to be a problem solving environment that supports the student in the task at hand, to perform an external representation of the system's expert and instructional models (see McTaggart 2001).

The interaction of all above-mentioned components must be implemented so as to provide the student with a face-to-face encounter with the knowledge of a subject domain. It is crucial that this internal *ITS* interaction act properly so that the student will be able to successfully assimilate new knowledge into his current mental schemata. There are different types of *ITSs*. It is beyond the aim of this book to provide an overview of them. The book will focus only on their common fundamental features.

Also the main problems and drawbacks that are typical to different kind of *ITSs* will be presented as a useful selection criterion of the different *CI/AI* techniques to use in improving the *ITSs* as a form of hybrid intelligent system.

Some of the main problems within the field of *ITS* will be mentioned in the following paragraph, see also (Negoita and Pritchard 2003b).

6.2.4 Drawbacks and Problems within the Field of ITS

Tractability and the Complexity Problem. No amount of computer power will be enough for an ideal "digit" teacher. Because of the problems of intractability brought by the complexity of the algorithms involved and the amount of computation to be made, Artificial Intelligence, in its initial stages, neglected decision theory and Bayesian probability (McTaggart 2001). Other methodologies were developed to replace *AI* in working on the previously mentioned problem; namely fuzzy sets (Zadeh 1983) and the Dempster-Shafter theory (Shafter 1986).

Scaleability Problem. Scaleability means a reduced number of constraints included in the knowledge database and/or actions that the system may want available. It can also mean thesizing to different problem domains. Scaleability becomes a problem because of intractability caused by complexity. As the systems size up, the complexity rapidly increases.

Lack of Reusable Components in ITSs. This drawback is regarding multiple aspects such as: the lack of standards of knowledge representation desirable for teaching, the lack of a standard interface to allow applications to access the knowledge and also the lack of a set of tools to allow designers to manipulate the knowledge.

Lack of Natural Language Capabilities in ITSs. Natural language is required for suitably connecting an *ITS* with mental schemata. More and more the understanding of human learning is being included in *ITSs* (McTaggart 2001), but this process is strongly limited by the lack of natural language capabilities of information technology that is actually used by these systems. The mind of each learner doesn't just process the information, but because of the interaction of the mental schemata with real life experience, it is actually created. An ideal *ITS* is desired to be able of achieving this connection with mental schemata through natural language, a dream impossible to be achieved nowadays, but perhaps only when a biological model of computing becomes a reality. It is definitely desirable that an ideal *ITS* be able to achieve this connection with mental schemata through natural language. At the moment this dream is impossible, but perhaps when a biological model of computing becomes a reality it will be the time for dreams to be fulfilled.

Not Meeting all Aspects of Learning. The social aspect of learning must some how be included within *ITSs* (McTaggart 2001), namely an integrated learning, including the *ITS*, that provides both agents that instruct and learn, and a collaboration between pupils. The agents could serve as personalized computer-based teachers, mentors and study guides for students, this being viewed as the ultimate humanization of computers (Baylor 2000).

Perceived Weakness in the Expert-centric and Efficiency-centric Approaches to Student Model Component. A main feature of recent *ITSs* is the Bayesian network student model of which three approaches are known: *Expert-centric* (Mayo et al. 2000), *Efficiency-centric* and *Data-centric*. Definite weaknesses were perceived in both the Expert-centric and Efficiency-centric student models, so that the use of a Data-centric student model was suggested in CAPIT ITS (Mayo and Mitrovic 2001).

The *Expert-centric* approach is developed in a manner similar to that of expert systems: the complete structure and conditional probabilities are determined, either directly or indirectly, by a domain expert and then every effort is aimed to fit the student model to the problem domain. The main disadvantage of this model is that the problem domain can become too large – there being too many variables to accommodate. So the whole system becomes intractable. It becomes infeasible to make a real-time adjustment of the student model and also in the absence of any data the conditional probabilities can't be defined. If these probabilities are defined beforehand by a domain expert, a problem appears with respect to these definitions that they must be appropriate when applied to the student model.

The *Efficiency-centric* approach is exactly opposite to the Expert-centric one: the student model realizes the complexity problem, being restricted or limited in someway and the problem domain is made to fit the resulting model. This restriction of the problem domain may have an undesirable result in the exclusion of some important or critical variables. The assumption seems to be that there is no dependence between variables, which in real life is highly unlikely. Loosing a critical variable may result in the collapse of the whole structure.

Compression of Variables within the Data-centric Approach to the Student Model. The Data-centric approach has a number of advantages over the Expert-centric and the Efficiency-centric approaches to the student model component (Mayo et al. 2000; Mayo and Mitrovic 2001). For example, because only observable data is used the student model is smaller and the model's predictive ability can be easily measured by testing the network on data that were not used to train it. But the model's drawback consists of the problem that the number of observable variables can still be high. Despite the possibility of grouping and compressing the variable, there is an issue of how this compression should be performed.

High Cost of Time Consuming Design Process. The biggest problem in building *ITSs* is the high cost involved in design. As *ITSs* become increasingly used and effective they are also become more difficult and expensive to develop (Murray 1999).

6.2.5 WITNeSS – A Fuzzy Neural Soft Computing System for Optimizing the Presentation of Learning Material to a Student

Intelligent Technologies either standalone or by *hybridization* overcome the weaknesses of classical *ITS*. *Fuzzy Systems* are removing the tractability/complexity problem and also offer the possibility of connecting an ITS with mental schemata using natural language. Designing *ITSs* with *Integrated NN* based learning environments provides agents that instruct and learn along with a student and his/her fellowpupils, helping to include even the social context of learning in the system. *Bayesian Networks* are used to reason in a principled manner about multiple pieces of evidence. The **WITNeSS** (**W**ellington **I**nstitute of **T**echnology **N**eural **E**xpert **S**tudent **S**upport) was designed as an hybrid intelligent system, a *Fuzzy – Neural Soft Computing system for optimizing the presentation of learning material to a student.*

WITNeSS Structure

Just the functionality of **WITNeSS** will be briefly described as illustrated in Fig. 6.2. A more detailed description of some component blocks may be found in (Negoita and Pritchard 2004a,b).

An interesting *HIS* aspect is to explain the hybridization flow of *FS* and *NN* technologies in *WITNeSS* (see Fig. 6.3). At the beginning of each block of work (learning), input data from the *Student's Data File* is feed into the *Optimized Fuzzy Rule Inference Machine*. The *Student Fuzzy Rule Profile* is loaded into the *Fuzzy Rule Inference Machine* when the student first logs on. Also, at varying times, a new finely tuned student fuzzy profile is loaded into

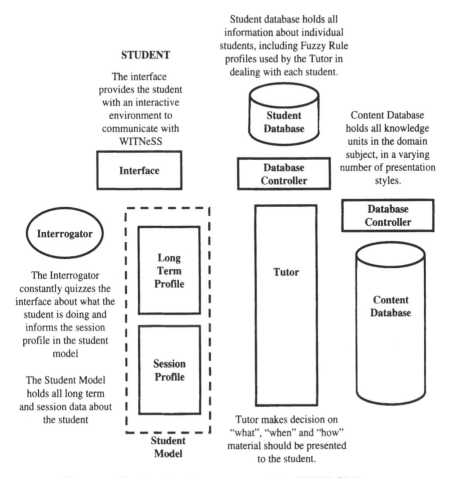

Fig. 6.2. The functional components of the WITNeSS System

the *Student Fuzzy Rule Profile* module. This is done when the current *Profile* is feed into the *NN*, finely adjusted by the *NN* and extracted using a fuzzy system to give the new set of refined rules for that student.

The *Fuzzy Rule Inference Machine* makes the decisionas to *"what"* and *"how"* to present the next block of learning content to the student. These intelligent decisions are presented to the *Content Database*. The *Content Database* delivers the content (to be learned) wrapped in a presentation package to the *Computer/Student Interface*, namely to the *Work Presented to the Student* block. The output of this last block together with the output of *Computer/Student Interface* (the students works on the learning activities presented) are inputs to *Student's Work Evaluated*. An evaluation of the student's work with this block of learning content is feed back into the *Student's Data File*. Some of these data will be feed into the *Fuzzy Rule Inference Machine* so that a decision can be made on the next block of learning (work). At

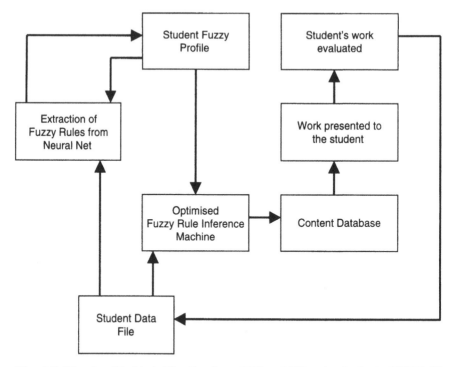

Fig. 6.3. The simplified hybridization flow of *FS* and *NN* technologies in *WITNeSS*

this point the learning loop is restarted, introducing another block of learning. Historical data from *Student's Data File* is permanently feed into the *NN* (Negoita and Pritchard 2003b).

Sequence of New Probability Dynamics for Getting a Problem Right

ITS performance must be evaluated so that a decision can be made whether to develop an original *ITS* or compare two different intelligent systems to see which is better. This evaluation must be based on the ability of the system to optimally present learning material to students so that the student experiences maximum learning.

The problem being addressed with the learning systems is that there is only so much learning material that can be presented in a daily session and how should the learning system decide on which learning material to present. Does it present new material, revise material already learnt or provide a mixture of the two? What section of content should be presented? The idea of a test is that each system would be tried out on students, the learning results recorded and analyzed for performance evaluation. A problem arose whereby it wouldn't be possible to organize human students for the experiment. This was due to real world time constraints and the time of year. The question became, would it be possible to create *"virtual" students* that could simulate certain human learning behavior. Using *HIS* technologies a student object was created that as close as possible imitates the behavior of a human student. It must be said, that no matter what concept of design is used, this model will be certainly limited in the way it imitates the behavior of the human student.

A *Student Object (SO)* was created that is a Java class designed to imitate certain human student learning behavior when it interacts with an intelligent learning system. Basically the student object can *learn* and *forget* – just like a human student. The key to the working of *SO* is the calculation of the probability that the student will get the next problem correct. It is the functionality of the *SO* that determines the probability for getting a problem right after learning has taken place. More details on behavioral structure of *SO* are in Fig. 6.4.

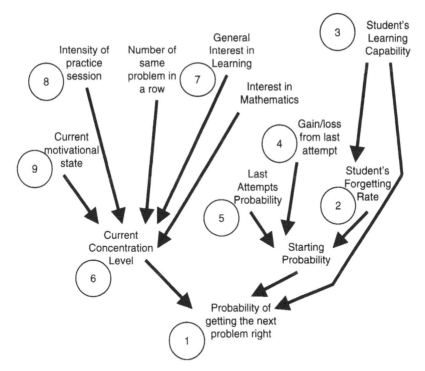

Fig. 6.4. Behavioural Structure of the "virtual student" – Student Object

The notation used in Fig. 6.4 is explained as follows:

1. *Probability of Getting the Next Problem Right.* The key to the working of the student object is the calculation of the probability that the student

will get the next problem correct (based on *Current Concentration Level, Starting Probability* and *Student's Learning Capability*).

2. The *Forgetting Rate* is used to increment/decrement the *Gain/Loss made from the Last Attempt*. This forgetting rate varies depending on the *Student's Learning Capability: HIGH* = 10%; *MEDIUM* = 20%; *LOW* = 40%. The assumption is made that the student with a high learning capability tends to forget less, while the student who has trouble learning tends to forget more.

3. The *Student's Learning Capability is* set when the *SO* is first created. This factor can be set at *HIGH, MEDIUM* or *LOW*. The assumption is made that the student with a high learning capability will tend to learn quickly and vice versa.

4. *Gain/loss from Last Attempt*. As a result of learning process, the student will either make a gain or loss in the probability of solving the next problem right. The gain/loss is adjusted by the *Student's Forgetting Rate* and then used with *Last Attempts Probability* to set the *Starting Probability*.

5. *Last Attempts Probability*. This is the probability level resulting from the last learning experience with this type of problem. It is the starting point of calculations.

6. *Current Concentration Level*. It doesn't matter how well the student's doing or what their learning capability is. It really depends on how well they can concentrate and focus at that moment. Like other factors (keeping it simple) the value is calculated as *HIGH, MEDIUM* or *LOW*.

7. *General Interest in Learning* and in particularly *in Mathematics* will result in greater and quicker learning of the content presented to the student. Both of these factors are given values of *HIGH, MEDIUM* or *LOW*. A *HIGH* value will tend to help learning the most.

8. *The Intensity of the Practice Session*. The student will react to the *Intensity of the practice session* and the *Number of Same Problem in a row* differently. The student with a *HIGH* learning capability will learn quicker with these factors *HIGH*, but a student with *LOW* learning capability will get bored quickly, and a *HIGH* in these factors will tend to hinder learning.

9. *Current Motivational State*. It reflects the instantaneous environmental changes in motivation. Students with *HIGH* learning capability will tend to be less affected by these changes.

A relevant image of WITNeSS working may be relied on Fig. 6.5.
The description boxes with regard to Fig. 6.5 are as follows:

1. When student logs on all information – long term profile, starting session profile, fuzzy decision rule profile – is loaded into the *Student Database*

2. Although the student starts with standardised profiles and rule sets, as the student interacts with the system, the system, using hybrid intelligent technologies, will fine tune these profiles and rules

Note: the numbers in brackets on the flow arrows refer to the description box
of the same number.

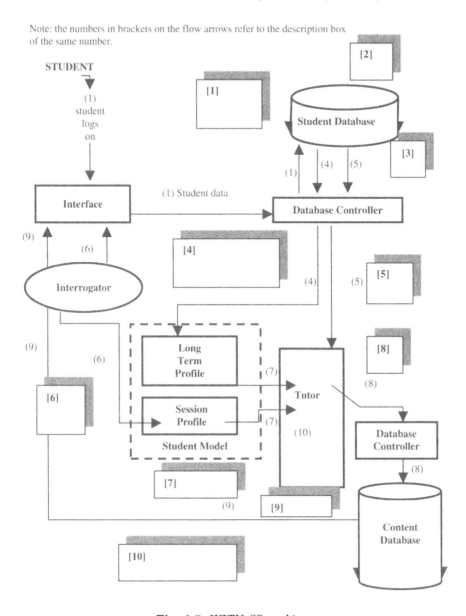

Fig. 6.5. WITNeSS working

3. What happens is that the system will learn how to provide the best
   tutoring service for a particular student. The students will end up with
   their own individualised profile and rule sets
4. Current information about the student plus their long-term profile is
   loaded into the *Student Model*

5. Student's profile of fuzzy rules for determining the next content and presentation mode is loaded into the *Tutor* module

6. The *Interrogator* starts to gather information from the interface about the student's behaviour. This is feed into the *Session Profile* module of the student model

7. Student current information, *Long Term Profile*, current session data is feed into the *Tutor* module

8. Using the fuzzy decision rule structure, Tutor determines what content modules to present to the student next

9. Content *Database Controller* arranges for the appropriate modules to be sent to the student interface

10. When the final data about the session comes back from the *Interface* via the *Interrogator* and *Student Model*, the *Tutor* module using its fuzzy-neural component – fine tunes the fuzzy rules in an effort to find a better what of making decisions on the student's behalf. The process now repeats from step [8] through to [10]

6.2.6 The Optimiser Agent in WITNeSS

The *Student Model* is placed into a typical *ITS* system configuration (McTaggart 2001), and an agent we call the *"Optimizer"* was added. The system, as a whole, works in the way presented in the previous section.

The *Student Model* is used with an optimizer agent to "fine-tune" the linguistic variables of a fuzzy rule decision structure that is used by a *Tutor* model to decide on *"what"* should be presented next to a student and *"how"* it should be presented. There is a detailed description of how the concept of an *"optimiser"* works, see (Pritchard and Negoita 2004).

The role of the *Optimizer* runs in the following way:

– The *human student* interacts with the system, which presents a sequence of learning activities that result in the student learning the knowledge structure. The key to this learning sequence is the fuzzy rule decision structure used by the *Tutor* agent to decide on *"what"* to present and *"how"* to present it. However the "shape" of the linguistic variables held in these fuzzy rules doesn't guarantee optimal learning. Each student learns differently, so what the *ITS* requires is the system to find the best "shape" for the linguistic variables, for that particular student. Hence the motivation for an *"Optimiser"* agent.

– When the student first logs onto the system, the *Optimiser* agent makes its own copy of the main agents (*student, tutor* and *knowledge*), including the current fuzzy rule structure. While the student is working with the main system, the *Optimiser* works in the background, trying different "shapes" to the linguistic variables in an effort to improve the current fuzzy rule structure. This would result in the tutor deciding on a quicker, higher quality learning sequence.

The *Optimiser* sends its "shape refined" linguistic variables back to the main Tutor, replacing its fuzzy rule structure with a more efficient one. The *Optimiser* then starts over, taking the latest information about how the student is performing, and works to find an even better "shape" to the linguistic variables. All the time the *Optimiser* is trying to come up with a better plan to take the student from where they are currently, to the end of the programme.

What the *Optimizer* is doing is illustrated in the following steps:

STEP 1 – The system creates the *Optimiser*, which in turn, creates copies of the *Student, Tutor* and *Knowledge* agents from the main system (see type 1 arrows in Fig. 6.6). It also copies the current fuzzy rule decision structure (see type 2 arrow in the same figure). It then creates an initial *GA* population of 20 chromosomes. Each chromosome represents changes that *can be made* – (*possibility be made*), to the shape of the basic linguistic variables (see type 3 arrows in Fig. 6.6).

STEP 2 – Each chromosome can have up to 10 genes in it. Each activated gene represents a change that *can be made* (possibly *be made*) to the shape of any of the linguistic variables (see *key* 4 in Fig. 6.6).

STEP 3 – A chromosome is passed to the *Tutor* and used to modify the "shapes" of the linguistic variables resulting in a different variation of the current fuzzy rule structure (see type 5 arrows in Fig. 6.6).

STEP 4 – The *Tutor* uses this modified version of its fuzzy rules to take learning activities from the knowledge agent and also to present them to the student (see type 6 arrows in Fig. 6.6).

STEP 5 – Both keys 6 operations are iteratively repeated until the student has learnt the whole knowledge structure (see key 7 in Fig. 6.6).

STEP 6 – The student is tested to see how well the knowledge has been learnt. The chromosome is evaluated using a Fitness Function of the following form (also see type 8 arrows in Fig. 6.6):

$$f(t, q) = t \times q$$

where:

$f(t, q)$ – fitness value of the chromosome.
t – number of steps taken to learn content.
Q – number of errors student makes in the test.

The population of chromosomes is ranked in a descending order by fitness value, the top 20 individuals are kept and the rest of them are discarded (see Fig. 6.6). Each new generation is operated on using *GA* operators of selection, crossover and mutation, to arrive at approximately 40 chromosomes. The whole process then goes back to *STEP 5* – *Chromosome passed to Tutor*. A number of generations are produced until the stop condition has been meet.

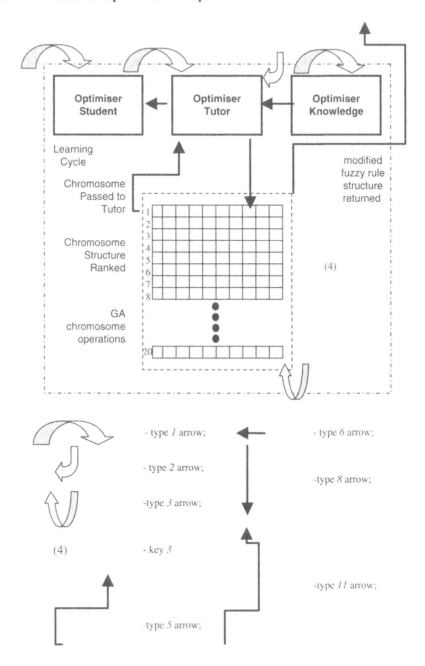

Fig. 6.6. How the Optimizer agent works

The best chromosome is passed back to the main *Tutor* agent where it is used to modify the "shapes" of the linguistic variables. This gives a more efficient fuzzy rule decision structure for this particular student (see type 11 arrow in Fig. 6.6).

Selection is done using the "roulette wheel" method and the parents retained, meaning that both parents and children must "compete" to stay in the population. Ten breeding couples are selected and the same chromosome may be selected more than once.

Breeding is done using a two-points crossover. Once the breeding process is finished, the *GA* undergoes mutation. Whereas the unit for operating the *GA* selection is the whole gene, in mutation it is the internal structure of the gene, which is being modified.

Two kinds of mutation occur. The first is used to try and allow for slight changes within a narrow volume in the solution space. It is an effort to encourage convergence onto a possible solution. Each gene within the chromosome has a slight probability to be chosen for mutation and only one node of the 4 nodes within the gene will possibly be mutated. This *type-A mutation* is only carried out on the parents involved in breeding. In the second type of mutation, increasing the probability of a gene being selected for mutation encourages a wide expansion of volume in the solution space and if this happens, all nodes of the linguistic variable may be changed. This *type-B mutation* is carried out on those chromosomes not selected as parents. No mutation is carried out on the children.

To allow for the fact volumes of the solution space not investigated might hold the optimal solution, five new chromosomes are created at the beginning of each new generation. If they turn out to be bad creations, they will simply be eliminated during the evaluation and sorting of the chromosome structure. If however they show promise, they will stay to "compete" with the other possibilities.

6.2.7 Concluding Remarks on WITNeSS

The *HIS* framework offers a new line of thinking in developing intelligent tutor systems or learning environments that can dynamically adapt its scheduling of teaching to result in quicker, more efficient tutoring of an individual student. The results of the experiments introduced in the previous subsections are supportive to above-mentioned remarks.

However there is still work to do in the future before *WITNeSS* comes close to a real-life intelligent tutoring system. The work must be more focused on details regarding the cost analysis of the system in terms of processing resources.

Experiments were conducted to test *WITNeSS's* ability to optimally decide on "which" learning material to present to students and "how", so that the students experience maximum learning. Each time *WITNeSS* was tried out on virtual students, the learning results were recorded and analysed, and a comparison against other learning systems made.

The problem arose whereby it wouldn't be possible to organize *"human" students* for the experiments. This was due to time constraints and the time of the year. The question became, to create *"virtual" students* that could simulate certain human learning behaviour, so the *Student Object* has been created. The concept of *"virtual" student* is a key concept for *ITS* testing and its further development.

Of high priority will be the improvement of how the system performs when confronted with a real-life situation – real students and a more complex knowledge structure. Some work has already started in developing a student model that represents more accurately the particular features of students working with the system. These improved knowledge structures and student models will, for sure, add greater efficiency to the system.

While *WITNeSS*, at the moment, is reliant on a *FS-GA* hybrid method, we still need to look at the possibility of other hybrid intelligent techniques being used in making this *ITS* more efficient. For example the efficiency of pattern recognition tasks may certainly be improved by using Artificial Immune Systems.

6.3 GA Relied Hybrid Intelligent Systems for Optimisation Objectives of Fuzzy Systems and Neural Networks

Real-world applications have proved two important facts: *FS, NN, EC* are complementary and not competitive *CI* techniques, so in most of the cases these methods are used in combination rather than exclusively. Each of the principal intelligent techniques has its own specific contribution with the aim of increasing the efficiency of human-like intelligence in non-living systems, namely:

- *FS* builds the *communication* function
- *NN* develops the function of *thinking*
- *EC mimics the evolution* as dictated by constraints of the external environment.

But *EC* has another functional role as well. It is an optimization technology to be of help both to *FS* and *NN* because these two techniques lack their own intrinsic optimization tools. This was another reason for applications development involving *GA* relied *HIS* for optimization objectives regarding *NN* and *FS*.

6.3.1 GA with Variable Length Chromosomes as an Optimisation Tool in Complex Hybrid Intelligent Systems

The methodological background and terminology of genetic algorithms – GA – as well as of all other evolutionary algorithms – EA – are rooted in the so called neo-Darwinian paradigm, whose elements were taken from Darwin's evolutionary theory, genetics and population genetics. The fundamental principles of these paradigms reflect the fact that natural organisms have evolved from simple to ever more complex ones with associated increase of genotype length.

Some features of traditional GA are leading to a recombination operator having a simple implementation: they use genotypes of predetermined length, that is homologues features are coded on homologous positions within the genotype.

Complex organisms in nature present a kind of *gene's specialization: structural genes* (**qualitative**) and *adjusting genes* (**quantitative**). The adjusting genes affect the structural genes globally (Zimmermann et al. 1996). Another interesting evolution aspect is that a correspondence to a recombination operator that involves individuals with significant different genotypes is to be found nowhere at any level of beings in the natural environment.

These two last aspects of evolution in nature made possible implementation of optimization methods in form of $VLGGA$, where:

– absolute position of the symbols usually does not provide information about the related feature;
– the recombination operator should be implemented in a different manner. Most implementations are application-dependent.

$VLGGA$ are successfully used in getting better performances for systems with a complex structure or, at the same performance level, a less complex structure of the system (Fagarasan and Negoita 1995). The flow-chart procedure of our $VLGGA$ is as from (Fagarasan and Negoita 1995), see Fig. 6.7:

STEP 1 – *Generate initial population* individual by individual (two fields for each of the individuals: *structural genes* gen eration, followed by *adjustable genes* generation);

STEP 2 – *Group* the individuals *in species*;

STEP 3 – *Evaluate* the initial population

STEP 4 – **repeat**
> **for** each *species*
>> advance *(selection, crossover, mutation)*
> **end**

STEP 5 – *global selection on the whole population*

STEP 6 – **until** *STOP conditions*

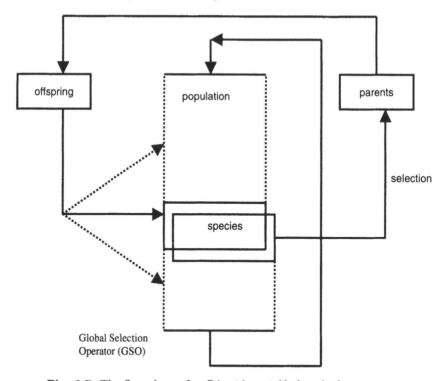

Fig. 6.7. The flow chart of a *GA with variable length chromosomes*

More details and explanations will be found for each of the above-mentioned steps in the following two Subsects. 6.4.2 and 6.4.3 – strictly connected to the referred applications.

6.3.2 VLGGA Based Learning the Parameters of a Fuzzy Inference System from Examples

Fuzzy Inference Systems (*FIS*) provide models for approximating continuous, real valued functions. The successful application of fuzzy reasoning models depends on a number of parameters, which are usually decided by a subjective methodology. Traditionally, fuzzy rule bases were constructed through knowledge acquisition from human experts.

Learning fuzzy rules from data begins by defining the language of the rule base: the terms that may be used, in the antecedents and the consequent of the rules. This is accomplished by decomposing the input and output domains into fuzzy regions. The subdivision into regions localizes the learning process within each region of the input space representing the antecedent of a rule. The learning algorithm proceeds then, with determining the entries in a fuzzy associative memory. The value assigned to an entry is used to determine the response when input occurs within that region.

Both the fuzzy partitions of the input and output spaces and the rule base were formerly decided in a subjective manner, by knowledge acquisition from human experts. However it is important that a change in a structure of a rule may locally alter the performances of a *FIS*. Moreover, a change in the partition of the input or output spaces may alter the performances in a more global and significant way.

A systematic procedure for the design and optimisation of a *FIS* is required. The interest for this topic was reflected in the number of papers concerning methods for learning/optimising the parameters of a *FIS*. However, most of them treat the different optimisation tasks in a separate manner. It has been shown (Park et al. 1994) that the performances of a fuzzy controller may be improved if the fuzzy reasoning model is supplemented by a genetic-based learning mechanism. The results have indicated that the best performance of the fuzzy control system was achieved when the spaces of both fuzzy membership functions and the fuzzy relation matrix was searched by a *GA*.

It still remains a source of subjectivity in deciding the number of fuzzy labels for both input and output universes of discourse. It is obvious that a smaller number of fuzzy labels determines a smaller number of fuzzy rules and, consequently it is possible to obtain a less complex fuzzy model and a smaller computational time for the inference.

The aim of this section is to describe the main aspects involved in developing a genetic-based flexible method able to learn the parameters of a *FIS* using a set of given examples. These parameters are: a minimal number of fuzzy labels, the shape and position of the corresponding membership functions, and the structure of a minimal number of fuzzy rules such as the inference error related to the set of examples to be below a given threshold.

It is proved in (Fagarasan and Negoita 1995) that a non-linear continuous function $f : [0,1] \to [0,1]$ can be better approximated by a number of line segments with variable sizes (Triangular Partition – *TP* case) than by the same number of segments but with uniform projections on the horizontal axis (Equidistant Triangular Partition – *TPE* case). A much better approximation may be achieved in the case of hyperbolic segments (Triangular Partition with Variable Length – TPh case). It means that it is also possible to get the same approximation error but with a smaller number of segments i.e. a smaller number of fuzzy labels. Furthermore, changes in the shape of the transfer function can be easily achieved by modifying the inference method (Yag et al. 1994) or the defuzzification method (Kiendl 1994). A small error in a certain application can be achieved by a *TPE*-based system if complex enough. But the *HIS* described in this section uses a learning algorithm intending to achieve a certain level of performances with as simple a system as possible.

The task of this *HIS* is to simultaneously learn the parameters of a *FIS*. For example, the number of fuzzy labels for inputs and output, the shape and position of the membership functions and the structure of the rule base. Because the chromosome code structures are of different complexities, the length of these chromosomes is subject to change during the algorithm. See also (Harvey 1992). Therefore the genetic operators and the strategy of evolution are adapted to these circumstances. The main features of the genetic-based learning method are as follows:

- the *chromosme* is structured by three fields: the *FL*-field that codes the number of fuzzy label for inputs/outputs; the MF-field that codes the shape and position for the fuzzy membership functions; the *FR*-field that codes the structure of the rule base
- the *initial population* is randomly generated as follows: first, the *FL*-field is randomly generated for each chromosome; then according to the number of fuzzy labels, the next two segments (fields) are generated; the phenotype corresponding to the *FL*-field is used to compute the length of each chromosome; the *FL*-field gives the complexity of each structure and the MF and FR fields give the structure itself.
- the *fitness function* is computed in order to combine two different optimization problems, i.e. a minimal error related to a set of given examples and a minimal structure (meaning a minimal number of fuzzy labels for input and output and a minimal number of rules in the rule base);

The fitness function comprises two main terms:

$$\text{Fitness (chromosome)} = \alpha E1 + \beta E2$$

where

$E1 = E1$ (error, threshold)
$E2 = E3$ (number of labels, number of rules)
α, β – parameters that give variable weights to each optimization task according to the application goals

The most important weight has to be for $E1$, but when the threshold is reached, the complex structures are penalized in order to minimize the number of label and rules.

The *GA* runs as described in Subsect. 6.3.1.

Each species contains structures of same complexity and advance independently. New species can be created when a mutation occurs in the *FL*-field, so the population size varies during the algorithm. When *GSO*-the global selection operator is applied, only the fittest chromosomes survive. In this way some species may grow and other disappear. The evolution is similar to that found in the natural environment where different competitive species share the same (limited) survival resources.

The *VLGGA* was tested on a two inputs/one output *FIS* that had to approximate a non-linear, real-valued function:

$$f : [0,1] \times [0,1] \rightarrow [0,1] \tag{6.1}$$
$$f(x,\ y) = 1/(1 + 3^*x - 1.5)^6 + (10^*y - 5)^6)$$

A set of examples were generated by using $f(x,y)$ and these examples (values) were used for attempting to achieve *FIS's* parameters so as to get a good approximation of $f(x,y)$. The *VLGGA* was set up in two ways:

- first, the best *FIS* approximation was searched for a *FIS* featured by *TPE* for the input space, equidistant crisp partition for the output space – 6 labels for each of them – and both *FL* and *ML* fields unchanged
- then, the modification of the *FL* and *MF* fields was achieved by using the best error as the threshold for the fitness function.

The evolution of the fittest chromosome in the population and it's corresponding total number of labels versus the number of *GA* iterations, led to the conclusion that in case of a *TP*-based system – variable length genotypes – the number of fuzzy labels first increases (complex structures get better approximations). But then the number of fuzzy labels decreases below the total number of labels in *TPE* case if the fitness function of the *TP*-based system comes near the given (penalty) threshold. This happens because the penalty function for the complex structure becomes significant. The mean square fitness function errors related to the same set of examples remain comparable both in case of *TPE*-based system and in case of the *TP*-based system. In conclusion it is possible to reduce the complexity of a *FIS*, the number of fuzzy labels and rules without altering the performance. This is possible if simultaneously both the fuzzy input/output partitions and the structure of the rule base are optimized. The proposed *HIS* also tries to avoid the difficult problem of recombination between parents with variable complexities.

6.3.3 VLGGA for Optimization of Fuzzy Recurrent NN

Various fuzzification approaches of neural networks (*NN*) have been proposed by using fuzzy numbers (Teodorescu 1994). A special class of hybrid intelligent (neuro-fuzzy) system, based on *NNs* and fuzzy algebraic systems was considered, for different *NN* topologies – multilayer perceptron, *R*adial *B*asis *F*unction (Kandel et al. 1998) and fuzzy recurrent *NN* (Arotaritei 2001). The application area of fuzzy *NN* is very large, from the prediction of time series to the modeling of complex chaotic systems. Most of the time the results are significantly better than those obtained in the case of crisp inputs, this justifying the increasing interest of researchers in developing more efficient learning algorithms for these types of intelligent hybrid structures.

The general learning methods for neuro-fuzzy *HIS* are based on gradient techniques, developed for working with fuzzy numbers. The well-known problems of gradient learning techniques (many local minima, learning factor determined experimentally, e.g.) are naturally present in these structures too, but their effect is greatly amplified due to constraint problems with regard to definition of the membership function of fuzzy numbers.

A recently proposed *NN* structure is a *Recurrent Artificial Neural Network* with *Fuzzy Numbers* (*RAFNN*), see (Arotaritei 2001). The recurrent topology is a fully connected one that uses symmetric triangular fuzzy numbers in the fuzzy algebraic framework. A computational expansive gradient-based algorithm was proposed and used in order to compute the matrix of fuzzy weights, but the structure (number of neurons) is supposed to be known, or discovered by simple incremental trials starting from minimal number of neurons set to two. *This can be very costly (excessive time consumption) in the case of modeling complex systems that could require large number of neurons.*

The gradient-based learning algorithm as mentioned above, was used to prove that a *RAFNN* could learn to solve an *XOR* type problem in the frame of the fuzzy numbers, continuously with two simultaneously changing inputs. This is not at all the case of a simple static *XOR* (with 4 possible input values and giving a typical *XOR* output to a *TTL* circuit). But the case is of a *RAFNN* learning the dynamics of a string of $0, 1$ that is randomly, continuously generated producing an *XOR* output that is delayed by 2–3 steps (in each update cycle, the teacher is delayed by $q = 2$ cycles relative to the input that is used for *XOR*, (Williams and Zipser 1989). The structure of a *RAFNN* (see Fig. 6.8), has n units and m inputs – an architecture that is similar to the crisp one in (Williams and Zipser 1989).

Each bias allocation in Fig. 6.8 will be seen as an input line whose value is always $(1, 1, 1)$. The set of indices for input units, output units and target units are denoted by (I, U, T). The inputs are denoted by $\tilde{x}(t)$, the outputs are denoted by $\tilde{y}(t)$ and $\tilde{d}(t)$ are the targets (if they exist) for each of the *RAFNN* units at time t. A generalized $\tilde{z}(t)$ was denoted similar to (Williams and Zipser 1989):

$$\tilde{z}_k = \begin{cases} \tilde{x}_k(t) & \text{if} \quad k \in I \\ \tilde{y}_k(t) & \text{if} \quad k \in U - T(t) \end{cases}$$

The *HIS* introduced in this section made use of a *VLGGA* implementing a *GA* learning algorithm both for the fuzzy weights of a *RAFNN* and for the *RAFNN* recurrent structure (number of neurons).

GA with variable length genotypes are typically used in getting better performances for systems with a complex structure or, at the same performance level, a less complex structure of the system (Fagarasan and Negoita 1995). Such a *GA* may be implemented and used for simultaneous optimization of both the structure and parameters of complex fuzzy inference systems (see Subsect. 6.4.2).

The genetic-based learning method used for simultaneously learning the *RAFNN* structure and parameters has the steps as described below.

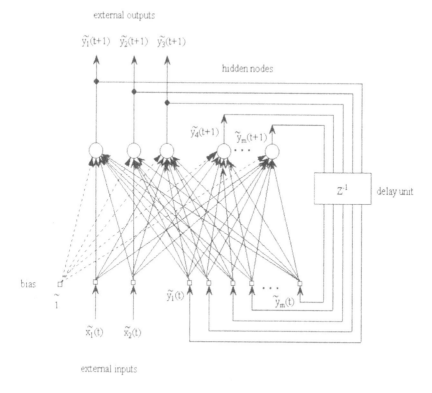

Fig. 6.8. The RAFNN Structure

STEP 1. The initial population is randomly generated in two steps for each of the individuals, as follows:

- first, the *structural genes* (representing n – the number of neurons) are generated
- these are random generated numbers, in a mapping format of 15 binary bits (0 or 1) corresponding to a number of neurons in a range $n \in [2, 15]$
- second, the *adjusting gene* is generated in form of a matrix with $n \times (m+n)$ elements
- m is the number of *NN* inputs, incremented by 1, the bias input ($m = 2$ in case of an *XOR* problem)
- the $n \times (m + n)$ matrix is a linear representation of fuzzy symmetric triangular numbers (see Fig. 6.9)

STEP 2. The individuals are grouped in *species* taking into account the geno-type-n – of the structural field in each chromosome, following an ascending order from the lowest to largest number n. A *species* means all the chromosomes with the *same length* (actually, in the broader sense, this means the *same length* and the *same structure*). This means more than the same length, because different individuals in the *VLGGA* may usually have the

STRUCTURAL GENE	ADJUSTING GENE
n genes (number of neurons in NN)	the fuzzy weight matrix n(m+n) genes

Fig. 6.9. The chromosome code structure for a *VLGGA* used in *RAFNN* optimization

same length despite having different structures. But *no special precautions are made with regard to this aspect in this application because of the suitable coding structure*. This is another advantage of using a *VLGGA* in this application.

STEP 3. Each individual of each species is evaluated during this step. The fitness function is defined in order to combine two different optimization problems, i.e. a minimal *NN* error related to a set of desired outputs and a minimal *NN* structure. The formula of fitness function g (*chromosome*) is as follows:

$$g(cromosome) = \lambda \cdot E_1 + \delta \cdot E_2$$
$$E_1 = |1.0 - J_{total}|$$
$$E_2 = 1/(1+n)$$

The experimentally determined parameters-λ and δ – show criteria weights in the composed fitness function. E_1 is the most important parameter and is given by the total error during the k-steps of tests (k is usually 1000). A penalty function (Coello 1999) is included in the fitness formula above.

The complex structure is penalized after p steps if the error J_{total} l has not decreased with a threshold δ (see Negoita and Arotaritei 2003):

$$\sigma = \sigma(k, \rho(J_{total}), \ generation)$$

STEP 4. Each species contains structures of the same complexity (see *STEP 2*) and advance independently under *common GA operators* (*selection, crossover, mutation*) as the *GA* are running in parallel. For species containing just one individual, only mutation is applied. The resulting offspring is appended to the species if it is better than the worst fitted chromosome in the whole species. In this manner each species grows as far as it produces well-fitted offspring.

New species can be created when a mutation occurs in the structural field, the chromosome lengths are modified, the individual goes to another species or another species is created. The number of species is subject to change during the algorithm.

STEP 5. After a predetermined number of iterations the *global selection operator, GSO*, is applied regardless of the species. The number of fittest chromosomes is fixed no matter what species they belong to.

GSO frequency application is by choice (application-dependent) and not necessarily at each iteration: species with short chromosomes are favored by a high frequency of *GSO*, while species with long chromosomes are favored by a low frequency of *GSO*. A *Boltzmann selection* was applied in this case with a probability as follows:

$$p = e^{-\frac{E_i(t)}{kT}}$$

where T depends on the iteration number *iter* and k on the chromosome length, $E_i(t)$

$$k = 1/(1 + l_{chrom})$$

$T(iter + 1) = 0.9.T(iter)$, $T(0) = T_0$ if *iter* is multiple of the predefined k_{gen}

$$E_i(t) = -J_{total}$$

Only the fittest chromosomes survive when *GSO* is applied: some species may then grow, others disappear. The evolution is similar to the natural environment where different competitive species share the limited resources.

In the work described in (Negoita and Arotaritei 2003), $T_0 = 17.5$. The initial population of 200 *NNs*, was split into 16 species; the number of members in each species ranging between 5 and 20. The label of one species is given by n – the number of neurons in the *RAFNN* and it belongs to integer numbers between 2 and 15. The population is generated using a flip coin probability. Relevant experimental results are introduced in (Negoita and Arotaritei 2003), regarding two aspects of *VLGGA* evolution during the *RAFNN* learning algorithm: the *VLGGA* fitness function and the optimized *RAFNN* output error.

A very good error for an *XOR* problem extended to fuzzy triangular numbers $Emax < 0.23$ – was achieved just after 4 iterations; this means the *RAFNN* dynamics were already stabilized. This stabilization was kept during the whole test, namely for 1000 iterations.

In the case of the *VLGGA* optimized learning algorithm, the optimal solution appeared only after 4 iterations and the constant behavior appeared after a magnitude order of 10^2. Meanwhile in case of the gradient method that was used in (Arotaritei 2001), the solution appeared after around 5×10^4 iterations. The improvement by more that one order of magnitude proved that a *VLGGA* reliant learning solution is significantly better than the gradient solution.

Usual, the test stage is considered successful as in (Williams and Zipser 1989), if the output of the selected neuron (in our case the last neuron of the *RAFNN*) produces a correct extended *XOR* delayed during a sufficient large number of steps. In our case the number of steps is 1000 and the delay is set to 2 steps.

Another interesting aspect, namely a typical one for *VLGGA*, was the evolution of the number of species in the population, along with the training

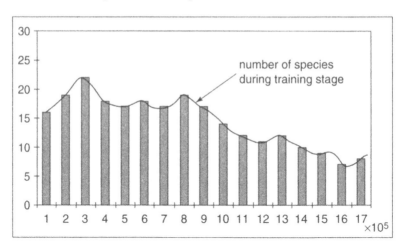

Fig. 6.10. The number of species evolution during the learning algorithm – the number of iterations is represented and on X axis and Y axis is representing the number of species

period of time, as presented in Fig. 6.10. The main remark regarding this figure: the $VLGGA$ moderates the pressure over a variety of individuals in order to avoid excessive elitism in species.

The experimental optimization results show a slow NN evolution to a point of maximum. There are a relative high number of required iterations (approx. 105) compared with the crisp case, but a real advantage in case of NN weights being a fuzzy number. A sensibly less number of iterations is required by the $VLGGA$ than by adapting gradient algorithm. The capability of a $VLGGA$ is illustrated by the fact that its performances were achieved in conditions of weight solutions in a form of (short) fuzzy numbers rather than crisp numbers. NN structures are usually determined experimentally in most of the applications, but this $VLGGA$ offers a systematic way of getting a minimal NN structure satisfying the requested performance. This advantage can not be ignored when a complex hybrid intelligent architecture must be designed without any previous details regarding its requested architecture (number of neurons and weight values in case of a NN).

The $VLGGA$ introduced in this chapter is different from (Harvey 1991) where chromosomes of variable length were considered, but the number of members of population is constant and the crossover is applied for members of the species regardless of the lengths of chromosomes.

The $VLGGA$ application area may be extended to other HIS using other NN architectures using non-symmetric triangular fuzzy number (three values coded for each fuzzy parameter). Another possible further extension could be made to other architectures using crisp number but also fuzzy numbers (i.e. multiplayer perceptron and RBF). The structure of the connection of neurons can be considered as a matrix of 1 and 0 values. A further application may include this parameter as the third encoding field in the solutions space. The first two fields in the actual *VLGGA* would be the same, the number of *NN* neurons and the matrix of fuzzy NN weights. This would mean that the connection gene field might be supplementary considered in the form of a matrix of connections, a matrix of 0_s and 1_s in the typical sense of classical notation in the graph representation by matrices of connection (Cormen 1990).

A more promising direction in research seems to be the *VLGGA* applicability to a *class of specific data mining problems* asking the question "which are the sequences that determine the behavior of the time series according to a defined objective?" Actually the response is known only for a given length of mined sequence, but a *VLGGA* reliant method would try to discover the length of the sequence also.

6.4 An Original Neuro-Fuzzy Knowledge-Based System for Air Quality Prediction: NEIKeS (Neural Explicit and Implicit Knowledge Based System)

The modular structures and methods presented in the previous chapters demonstrate the suitability of applications of connectionist systems in symbolic processing. Since the presented EKMs were demonstrated to be identical to Discrete Fuzzy Rule-based Systems (Buckley and Hayashi 1995), the homogenous integration of explicit rules and training data sets permits better cover of the problem domain.

In such cases the constraint of the size of neural networks is solved by the modularity paradigm. EKMs represent explicit rules identified by an expert or refined from IKM structures. IKMs are useful especially for complex problems described by (noisy) data sets. The EKM and IKM combination encourages compact solutions for problems described by both data sets distributed in compact domains in hyperspace, and isolated data situated in intersection of compact sub-domains or inhomogeneous intervals. After training, different expert networks compute different functions mapping different regions of the input space.

The different sources of the information explicitly and/or implicitly integrated in the expert modules exhibit the problem of knowledge redundancy in the final structure. The proposed methods, based on explicit and implicit module integration, can operate with redundant knowledge spread in both ways of representation, as discussed in Chap. 5. This redundancy should be clearly minimized. The method to implement redundancy minimization is based on selecting specific data sets from the training collection which are not suitable to verify the implemented rules. The resulted training data sets describe such domains in the hyperspace which are not covered by the explicit rules. The main disadvantage is that the IKMs are able to generalize just in their domain of expertise.

6.4.1 About Air Pollution Prediction

The methodology described in the previous sections was applied to the prediction of air pollution levels in Athens, Greece. Photochemical air pollution is a major environmental problem in contemporary large cities. This is a phenomenon involving a series of chemical reactions triggered by solar radiation. Non-Methane Organic Compounds (NMOC), nitric oxide and nitrogen dioxides (NO_x), ozone (O_3), and the energy that comes through a specific range of the spectrum of solar radiation are elements of the photochemical reaction. The photochemical pollutants, especially NO_2 and O_3, have significantly increased in the last decades, therefore environmental experts need to predict the future behavior of these pollutants.

This specific problem is three-dimensional and refers how pollutants concentrations are evolving in space and time. A similar problem is related to SO_2 and CO air pollution around large thermal power plants.

Athens, the Greek capital, has a long history of air pollution episodes. It is an urban area, surrounded by mountains, except from a southern wayout that leads to the sea. Photochemical pollution has significantly increased during the last few years, so there is a growing concern by the authorities with how to manage the air pollution. At PERPA, the Greek air quality monitoring organization, for the specific problem described above two main predictive tasks are sought: the NO_2 and O_3 daily peak. The predictive process performed by human experts takes into account the ultimate meteorological data, as well as the measured values of NO, O_3 and NO_2. Therefore, an efficient short-term air pollution prediction model would help the air quality monitoring organization to survey and propose emergency measures. These restraining measures are important to reduce the power of plants or traffic density in the area. The predicted value of pollutants is important because the estimation of pollution in an area where there are no measurements allows preventive action to be taken. These estimations can be distributed to local authorities and the local radio and TV stations so that announcements can be made to the public.

6.4.2 Related Work

Until now, several studies have been made and research papers have been published discussing the role the artificial intelligence tools could play in predicting photochemical pollution. In (Lee 1995), prediction of the atmospheric ozone concentrations using neural networks was proposed. A machine learning approach was used in (Sucar et al. 1997) to predict ozone pollution in Mexico City, and also in (Avouris 1995), applied in Athens. Studies on the usage of neural networks for short-term air pollution have been presented in (Boznar 1997; Sucar et al. 1997; Neagu et al. 2002).

A comparative study (Avouris and Kalapanidas 1997) has been done using three artificial intelligence algorithms: a case-based reasoning adapted algorithm (based on retrieval from the database of the most similar past cases subsequently adapted to the present conditions in order to provide the new solution), a fully connected multi-layer perceptron (MLP), and a common inductive decision tree approach using the Oblique Classifier (OC1, implemented by Murthy et al. 1994).

Related work examples dealing with the daily maximum temperature forecast used machine learning (Abdel and Elhadidy 1996), and combined sets of meteorological, social and industrial factors (Lekkas et al. 1994). In all cases, the neural network approach of photochemical pollution is restricted to feedforward architectures. Moreover, given the fact that dangerous peak values of pollutants (high and medium pollution) are rare in the data sets compared with the number of low pollution cases, all the studies cited above give good results just for low values. The model's performance deteriorates in the case of peak and medium values.

The failure in predicting especially medium values of the pollutant by all the models presented in (Avouris and Kalapanidas 1997) is due to the fact that the algorithms cannot distinguish clearly low-level cases from the medium and higher ones. It has been observed that higher the pollutant value, the rarer their presence in the data set. This is an important reason to develop a neurosymbolic architecture based on specialized modules in order to combine the experience acquired from learning data sets and the explicit fuzzy rules given by human experts.

6.4.3 Data Preparation

The data set used in the application is that described by (Avouris and Kalapanidas 1997) as specific to the air pollution monitoring station Patission of the Greek air quality monitoring organization in Athens (PERPA). Additional data has been provided by the Greek National Meteorological Service, called EMY, and from the National Observatory of Athens. The inputs relate to ultimate meteorological data (rain, inversion, solar radiation, wind direction, wind speed) and pollutant concentrations (NO, O_3, NO_2). These factors were computed according to the heuristic functions proposed in [2][3], representing the contribution of each feature to the NO_2 episode evolution. The measurements are averaged every hour at the monitoring station. The values are collected from 1 to 10 a.m. in order to predict the NO_2 peak value for the rest of the day.

In order to test the applicability of the neurosymbolic approach to the photochemical pollution problem, we ought to gather and process large data sets first. The cleansing of the data has been a necessary step towards high efficiency, since there is always a high cost for noise or uncertainty in the data sets. The errors in the input data can be classified into the measurement and sampling errors (Zickus 1999). Systematic measurement errors are caused by faults in the measurement instruments. Systematic sampling errors occur due to THE influence of specific local micrometeorological conditions. Example of systematic sampling error can be the wind speed alteration due to the contribution of local terrain frictional forces or higher pollutant levels due to closer emission sources. However, in the present study we were more interested in changes in parameter values than in absolute levels, since this type of systematic error does not play an important role in data analysis. Measurement uncertainties, as quoted by the device manufacturers, are about 5% of the permanent measurement value.

Another factor that influences air pollution is the random emissions due to the unpredictability of human activities. Using normalized and fuzzified inputs and outputs and neural representation assures a minimization of the effect of these errors through the entire data set. The pollution database studied in this chapter covers a two-year period (1990 and 1991). In order to construct a useful dataset from this database, we decided to include in the final data set only the features that were more relevant to NO_2 values. The specific fuzzy sets were proposed in a form as appropriate to technical vocabulary of human experts as possible. We will explain now how each factor has been represented and computed.

The Meteorological Factors

The meteorological factors were computed according to how "favorably" each feature contributed to the NO_2 episode evolution. Each factor has been designed taking into account the results of the statistical analysis on each correspondent attribute, and advice from field experts. Hence, the meteorological parameters influencing pollutant dispersion are the recorded precipitation and solar radiation levels (gathered by the National Observatory of Athens), temperature inversion below 150 m over the surface, rain factor (from Greek national meteorological service, called EMY), wind speed and direction (recorded by PERPA). The data sets for the meteorological factors were normalized before processing, by dividing their values with each factor's maximum value as described in Table 6.1.

Table 6.1. The normalization values for meteorological factors

RainToday	InversionToday	SolRadAt13	WindDirect	WindSpeed
1	3	4	360	12

There are two sources of data concerning the rain factor: one is the EMY prognostic bulletin and the second is the National Observatory. Both produce a binary-valued rain factor. As for the later, we had two precipitation measurements at our disposal: the duration and the amount of precipitation. The rule for the rain factor was computed (see Table 6.2) in the following way:

Table 6.2. The values of the rain factor

RainToday Factor	Nil	Possible	Light	Snow	Rain	Periodical	Storm
	Precipitation Prediction						
0	×	×	×	×			
1					×	×	×

IF RainHeight >0.5 THEN RainToday=1 ELSE RainToday=0,
where 1 stands for rain, 0 stands for lack of rain.

The temperature inversion plays a favorable role in the evolution of pollution, acting like a trap for the pollutants and blocking them into the lower part of the atmosphere. This phenomenon causes high concentrations of the photochemical pollutants, mainly when no winds are blowing at the same time. It is more favorable to cause an episode, as it contributes to higher concentrations near the ground. The temperature inversion factor is a function (Avouris 1995) depending on the temperature difference and the height where this difference occurs (Table 6.3). It produces a crisp value set.

Table 6.3. The values of InversionToday linguistic variable

Heuristic Inversion Factor (InversionToday)	TD (temperature difference) H (height from ground)
L0	No inversion
L1	TD $= 0$, H at 50 m
L2	TD > 0, H > 50 m
L3	TD > 0, H > 50 m, multiple inversion layers up to 1500 m

The mean wind speed in Athens during the period of analysis was 2.07 m/s, while the maximum value was 10.97 m/s. The strongest winds blew from the Northeast (Fig. 6.11), while the prevailing wind direction was from the Southwest and Northeast sectors (Fig. 6.12).

In fact, due to the geography of the Athens basin, the winds coming from the South are more favorable towards a NO_2 episode. Temperature inversion is due to the lower temperature of these winds. The wind predictions provided by EMY are categorical values defining the approximate direction and speed of winds, as they are evolving in the period of the corresponding weather bulletin.

Consequently, the linguistic variables considered for meteorological inputs are characterized by the term sets:

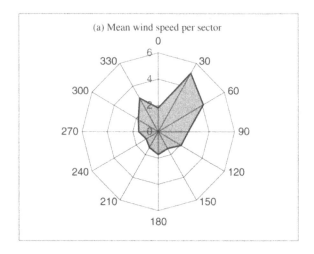

Fig. 6.11. Mean wind speed per sector

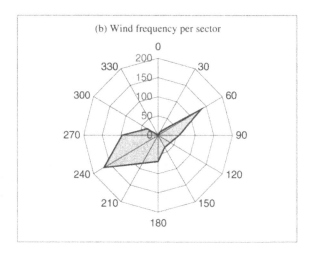

Fig. 6.12. Wind frequency per sector

RainToday={YES, NO},
InversionToday={L0, L1, L2, L3},
SolarRadiationAt13={LOW, MED, HIGH},
WindDirection={NE, S, NV},
WindSpeed={LOW, MED, HIGH}.

The fuzzy shapes of the normalized values of wind direction and wind speed factors, considered as linguistic variables, are presented in Fig. 6.13 and Fig. 6.14.

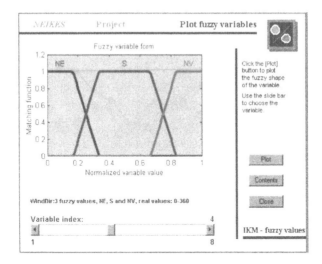

Fig. 6.13. Linguistic values for WindDirection

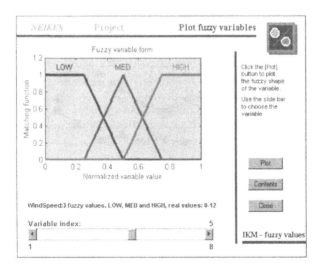

Fig. 6.14. Linguistic values for WindSpeed

The Pollutant Factors

Nitrogen oxides in the human respiratory system cause increases in both the susceptibility and severity of lung infections and asthma. Long-term exposure can also weaken the effectiveness of the lung's defenses against bacterial infection (WHO 1987). Nitrogen oxides are an important precursor to ozone and acidic precipitation, both of which harm terrestrial and aquatic ecosystems (Finlayson-Pitts and Pitts 1986). The nitrogen oxides and ozone take part in the photochemical transformations in the urban atmosphere, therefore it is

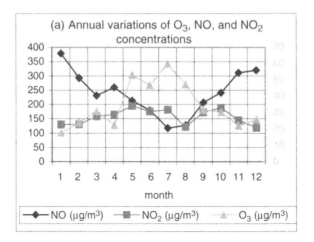

Fig. 6.15. NO, NO₂, and O₃ dynamics: annual variations (1990)

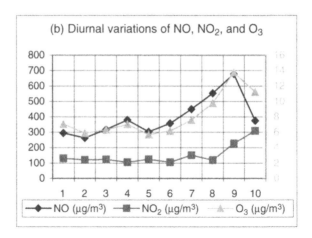

Fig. 6.16. NO, NO₂, and O₃ dynamics: diurnal variations (06/04/1990)

necessary to examine their dynamics together (Fig. 6.15 and Fig. 6.16). Nitrogen oxides NO_x are released into the atmosphere mainly in the form of nitric oxide, as a product of the reaction of NO with O_2 during high temperature combustion processes of the fossil fuel.

PERPA has been the main source of pollution data for Athens, as this public organization maintains an air quality-monitoring network consisting of 12 monitoring stations, equally distributed throughout the area. Four levels of concentration are defined for NO_2: level 1 (0–$200\,\mu g/m^3$), level 2 (200–$350\,\mu g/m^3$), level 3 (350–$500\,\mu g/m^3$), and level 4 (500–$700\,\mu g/m^3$).

Both the input data set values and the output ones were normalized and fuzzified with respect of the maximum observed values of pollutant concentrations. The maximum values for pollutants were considered: $700\ \mu g/m^3$

(NO_2), 1300 µg/m^3 (NO), and 130 µg/m^3 (O_3). The fuzzy shapes of the normalized values of the pollutants considered as linguistic variables are presented in Fig. 6.17 and Fig. 6.18, while their term sets are:
NO_2={L1, L2, L3, L4},
NO={LOW, MED, HIGH},
O_3={LOW, MED, HIGH}.

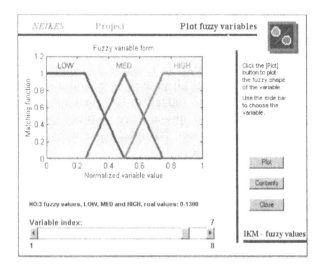

Fig. 6.17. Fuzzy values of linguistic variable NO

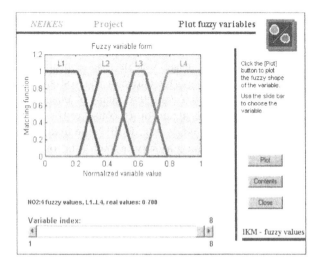

Fig. 6.18. Fuzzy values of linguistic variable NO_2

6.4.4 The Neuro-Fuzzy Knowledge-Based System
for Air Quality Prediction for Air Quality Prediction: NEIKES

The idea of using a neuro-fuzzy approach as a tool for air pollution prediction came from both the neural networks' ability to learn and generalize
from sets of historical patterns, and the complexity of the problem. Problems
that were more suitable to be modeled using fuzzy techniques (Fuller 1999).
Also, given the fact that human experts are already using some empirical
rules with imprecise and linguistic forms, a separate module is included with
rules as specific structured neural networks. The neuro-fuzzy networks, with
fuzzy inputs and output, are equivalent to the rule set, assuring a homogenous structure model as described in Chap. 3. Depending on the methods
used to combine the outputs of all processing modules in order to predict
the overall output of the system (NO2after10), three different structures of
the neurosymbolic system-based approach are proposed and discussed as the
main integration methods to build NEIKES (Neural Explicit and Implicit
Knowledge-based Expert System).

Data Base Partitioning

A research database was extracted for the period 1 January 1990 to 31
December 1991 (Avouris and Kalapanidas 1997; Neagu et al. 2002). The
whole set of available patterns was divided in two independent sub-sets, each
one of them having its own task in the model training and testing processes
respectively. A pattern is defined as a vector compound of values of the input features (selected pollution and meteorological parameters) and values
of the output feature. The training set of data was used for the adjustment
of the connections of both a MLP structure and, after fuzzification, a neuro-
fuzzy network based on the backpropagation algorithm. The testing set of
data was used for periodical testing of the implicit knowledge modules (the
trained MLP-based module and neuro-fuzzy networks) and explicit knowledge modules. The same testing set was used as a production set of data
(Boznar 1997) for studying the performance of the system.

The data set (730 cases, two complete years of observations and measures) was divided paying great attention to conserve the distribution of the
4 classes of linguistic variable NO_2 (Fig. 6.19). The algorithm was a 70–30
partitioning, as it is used in the majority of such kind of comparative tests
between predictive algorithms (Fig. 6.20). In such a situation the training set
contained 480 cases and the testing set 250 cases.

The Implicit Knowledge Structure

The implicit knowledge structure includes a MAPI-based HNN and MLP-
based network as described in Chap. 3, and a second IKM component integrated into SGN-based architecture. The first structure is a three-layered

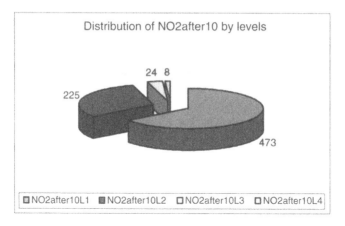

Fig. 6.19. The distribution of used patterns by NO_2 levels (levels L1, L2, L3, L4)

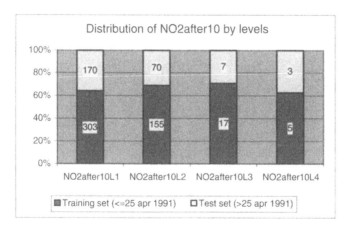

Fig. 6.20. The training and testing sets distribution of NO_2 by levels

network with the eight above described fuzzy inputs and one fuzzy output, the predicted value of NO_2 pollutant. The number of hidden neurons parameterizes the neuro-fuzzy network structure. The backpropagation algorithm (Rumelhart and McClelland 1986) was used to train both structures. A learning rate of 0.7 and a momentum term of 0.9 were used (a relatively high learning rate ensures rapid finding of the error function minimum and a high momentum term prevents too many oscillations of the error function). All the networks were trained for 1000 epochs, giving an error of around 0.005 for the MLP-based structure and 0.05 for the HNN-based structure.

Rules Extraction from HNN-based IKM

For the HNN-based IKM, the effect measure method combines the weights between the layers of the network in order to select the strongest dependencies between the output and the inputs (see Fig. 6.21). This approach takes advantage of the predictive capabilities of neural networks and gives reasons to explain the output and the patterns discovered by the IKM part of the proposed system. As well it highlights and adjusts some explicit rules given by human experts.

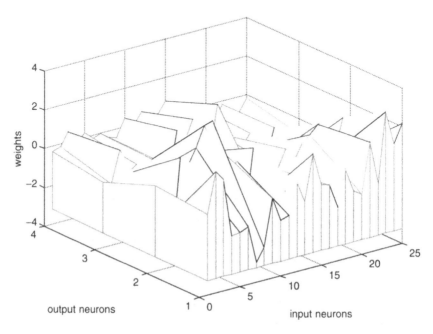

Fig. 6.21. Effect measure method: effect measure matrix for HNN with eight hidden neurons

After an analysis and interpretation of the conflicts between extracted rules, a set of rules, which constituted core knowledge about the problem, was determined. The strongest rules (average strength >30%) from such a set are shown below:

```
IF RainToday is YES      THEN NO2After10 is L1 (74.80%)

IF InversionToday is L0 THEN NO2After10 is L1 (58.95%)
IF InversionToday is L1 THEN NO2After10 is L2 (50.76%)
IF InversionToday is L2 THEN NO2After10 is L2 (33.32%)
IF InversionToday is L3 THEN NO2After10 is L2 (61.16%)

IF SolRadAt13 is LOW     THEN NO2After10 is L2 (95.10%)
IF SolRadAt13 is MED     THEN NO2After10 is L2 (100.00%)
IF SolRadAt13 is HIGH    THEN NO2After10 is L2 (45.70%)
```

```
IF WindDirect_10 is NV  THEN NO2After10 is L3 (35.56%)
IF WindDirect_10 is NE  THEN NO2After10 is L3 (47.47%)


IF WindSpeed_10 is LOW  THEN NO2After10 is L2 (34.07%)
IF WindSpeed_10 is MED  THEN NO2After10 is L1 (61.32%)
IF WindSpeed_10 is HIGH THEN NO2After10 is L3 (60.81%)


IF NO_10 is MED         THEN NO2After10 is L2 (31.54%)
IF NO_10 is HIGH        THEN NO2After10 is L1 (54.82%)
IF NO_10 is HIGH        THEN NO2After10 is L2 (44.99%)


IF NO2_10 is L2         THEN NO2After10 is L1 (99.40%)
IF NO2_10 is L3         THEN NO2After10 is L1 (41.75%)
IF NO2_10 is L3         THEN NO2After10 is L2 (45.33%)
IF NO2_10 is L4         THEN NO2After10 is L1 (73.23%)
```

Rules Extraction from MLP-based IKM

Following the steps described in the third integration strategy, the process of extracting acquired knowledge from the MLP-based IKM structure is proposed in Chap. 4. Obviously, the higher the number of hidden neurons, the complexity of the equivalent rules set increases. For example, for a network with 8 inputs, 4 hidden neurons and the NO_2 output were extracted four rules equivalent with the trained module, on the basis of *interactive*-OR operator:

```
IF (RainToday is not greater than approximately 60.16)
iOR(InversionToday is not greater than approximately 7.365)
iOR(SolRadAt13 is not greater than approximately 9.742)
iOR(WindDirect_10 is greater than approximately 4461)
iOR(WindSpeed_10 is greater than approximately 107.3)
iOR(O3_10 is greater than approximately 338.6)
iOR(NO_10 is greater than approximately 9352)
iOR(NO2_10 is greater than approximately 1.154e + 004)
THEN NO2After10 = 564.7


IF (RainToday is greater than approximately 5.664)
iOR(InversionToday is greater than approximately 11.45)
iOR(SolRadAt13 is greater than approximately 17.09)
iOR(WindDirect_10 is not greater than approximately 4364)
iOR(WindSpeed_10 is greater than approximately 83.6)
iOR(O3_10 is not greater than approximately 1.529e + 004)
iOR(NO_10 is greater than approximately 3181)
iOR(NO2_10 is not greater than approximately 4556)
THEN NO2After10 = 9.21
```

```
IF (RainToday is greater than approximately 10.34)
iOR(InversionToday is greater than approximately 6.896)
iOR(SolRadAt13 is not greater than approximately 9.735)
iOR(WindDirect_10 is not greater than approximately 1586)
iOR(WindSpeed_10 is not greater than approximately 165.8)
iOR(O3_10 is not greater than approximately 5429)
iOR(NO_10 is not greater than approximately 3079)
iOR(NO2_10 is not greater than approximately 1577)
THEN NO2After10 = -180.7
```

```
IF (RainToday is not greater than approximately 8.884)
iOR(InversionToday is not greater than approximately 6.559)
iOR(SolRadAt13 is not greater than approximately 36.42)
iOR(WindDirect_10 is greater than approximately 1484)
iOR(WindSpeed_10 is not greater than approximately 601)
iOR(O3_10 is greater than approximately 451.2)
iOR(NO_10 is greater than approximately 5548)
iOR(NO2_10 is not greater than approximately 2488)
THEN NO2After10= 991.3
```

The process of prediction for a given input is determined by an aggregation computation using the interactive operator (Benitez et al. 1996). The instance is matched against the rule premises, each rule being fired to a certain degree $v_i, i = 1, \ldots, 4$. The global output is the weighted sum of these degrees:
$Y=+564.7v_1 +9.21v_2 -180.7v_3 +991.3v_4$.

The Explicit Knowledge Structure

Seven rules, acquired from human experts, are represented as a HNN (as described in Chap. 3). These rules were expressed in a fuzzy form and cover the output domain (Neagu et al. 1999), (Neagu and Bumbaru 1999; Neagu and Palade 2000a) for premises based on the eight inputs. The EKM is build as a compact structure, following the method described in Subsect. 3.2.2, to be used in FEM or UGN integration procedures. The structure of EKM is based on the stand-alone $EKM_i, i = 1, 2, \ldots, 7$, built on the basis of Fire Each Rule method (Subsect. 3.2.2). For example, the rule number 4 is:

```
IF (WinDirect_10 is S)
AND(WinSpeed is LOW)
AND(O3_10 is HIG)
AND(NO2_10 is L4)
THEN (NO2After10 is L4)
```

while the rule number 3 is:

```
IF (WindSpeed_10 is LOW)
AND(NO2_10 is L3)
THEN (NO2After10 is L3)
```

and the rule number 6 is:

```
IF (RainToday is YES)
AND(WindDirect_10 is S)
AND(WinSpeed_10 is LOW)
AND(NO2_10 is L2)
THEN (NO2After10 is L3).
```

6.4.5 Results

Since the experiment studied the daily NO_2 maximum concentration prediction for the same day, the predicted outputs are indeed better than the
stand-alone modules done individually. A typical model performance is shown
in Fig. 6.22. Only about 5% of the days at the measuring station had highest
concentrations of the NO_2 pollutant. However, because they are very high
(levels L3 and L4), they are also very harmful to people and vegetation. For
this reason, it is very important to predict very high concentrations properly and not to make false alarms or optimistic predictions (the common
case specific to most of the machine learning approaches). The L1 level values are well predicted by each tool, including cased-based reasoning, decision
trees, or trained crisp values neural networks (Avouris and Kalapanidas 1997;
Boznar 1997). These tools include the implicit knowledge module used in the
current application. Combining both implicit and explicit knowledge using
fuzzy inference or gating networks is a promising step to better prediction of
high concentrations, without deteriorating performance for low values.

The performances of the models were evaluated on the testing data set
(250 cases), covering the period of years 1990 and 1991. The gating networks

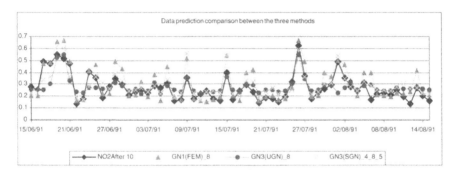

Fig. 6.22. Typical results of the NO_2 short-term prediction model

(UGN and SGN) are trained with 1000 epochs, this process assuring a reasonable behavior during the prediction process.

While the methods using gating networks are based on acquiring knowledge about the output distribution through the implicit and explicit knowledge modules during a training process (as described in Sect. 2.4), the first proposed method (FEM) computed the overall output using an inference step, as described in Figs. 6.23–6.25 (for a given testing pattern).

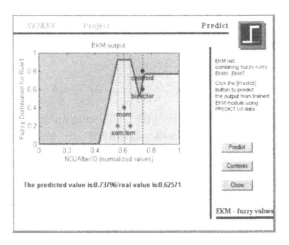

Fig. 6.23. The output of the Explicit Knowledge Module for the given input data set

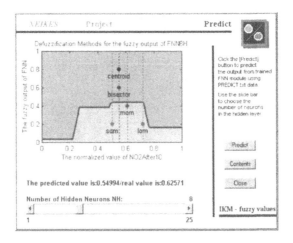

Fig. 6.24. The output of the Implicit Knowledge Module (FNN) for the given input data set

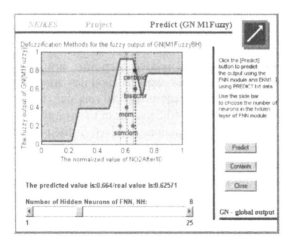

Fig. 6.25. The overall output of the system, computed as a T-conorm of the inferred outputs of the explicit and implicit modules of the system

One of the advantages of NEIKES is that, following the proposed methodologies, the behavior of all involved networks can be improved at any moment, based on new training data, choosing the number of hidden network or altering the explicit knowledge module starting from relevant rules extracted from the IKM part of the system. For this purpose, the user of NEIKES has the possibility to make a comparison between the predictions of all involved modules, for a given input set (Fig. 6.26).

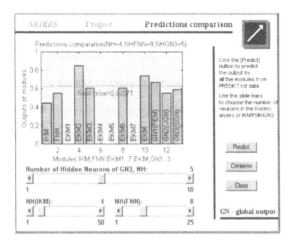

Fig. 6.26. Comparison between predictions of all involved modules for a given input data set

The difficulties in predicting the second and third level NO_2 peaks by other machine learning models (Lekkas et al. 1994; Avouris and Kalapanidas 1997) are tackled by our approach: while the ambiguous cases caused by very similar circumstances and related attributes are distributed into the implicit knowledge modules, the explicit modules adapt the shape of the overall output. The different methods used to combine the results inferred by each module determine a weighted behavior of the overall output, in the sense of taking into account all the opinions involved in the prediction process.

The most difficult task of such a system is to indicate decisively whether the current day is a NO_2 episode day or not. According to the air pollution experts, this first decision is necessary to drive the subsequent phase of counter-measures proposal, i.e. issue of public warnings, traffic control measures, etc. In air pollution prediction, the significant L3 and L4 levels of the pollutant are represented by a small number of training data. Consequently, the explicit rules, given by experts, have the role to improve the overall performance of the proposed system, as seen in Fig. 6.24 (a consequence is the absolute difference variation dependent on the architecture between computed output and real value, as exhibited in Table 6.4). Moreover, the possibilities to improve both, the training modules and the EKMs through a knowledge acquisition process underline the adaptability of such architectures, while the results can be represented in fuzzy forms, to give some semantic explanations to the user. The statistical results, as depicted in Figs. 6.27 and 6.28, have to be carefully interpreted for high values of the output (levels L3 and L4), while the testing values represent just a small part of the entire data set (see Tables 6.5 and 6.6). As an example, the third L4 episode of our testing data set, unpredicted by all the structures, is recognized, in fact, as an L3 episode by all the proposed architectures, a good starting point for alarm (this value is the maximum value in the entire data set, corresponding to the episode of October 1st, 1991).

Table 6.4. Maximum value of absolute difference between computed output and real value

NO_2 Level	IKM(MLP)	IKM(FNN)	GN1(FEN)_8	GN2(UGN)_8	GN3(SGN)4_8_5
L1	0.193155	0.258925	0.324865	0.270351	0.203363
L2	0.179727	0.204155	0.212101	0.273066	0.207605
L3	0.329247	0.277584	0.210021	0.412509	0.31933
L4	0.459402	0.404002	0.391603	0.509165	0.379941

We assume that a larger data set (preferably five to ten recent years cases) would present a better level distribution, more accurate to the actual population that would increase the possibilities of the methods to predict

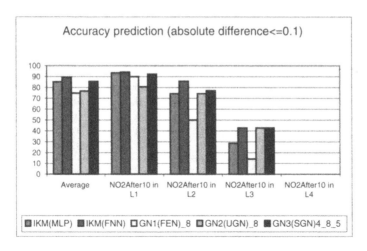

Fig. 6.27. Accuracy comparison (%) for an absolute difference of 0.1 between the computed output and the testing value

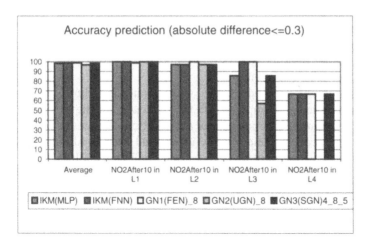

Fig. 6.28. Accuracy comparison (%) for an absolute difference of 0.3 between the computed output and the testing value

Table 6.5. Exact IKM prediction accuracy and failures by levels

NO_2 Level	IKM(MLP)				IKM(FNN)			
	L1	L2	L3	L4	L1	L2	L3	L4
L1	150	20	0	0	157	13	0	0
L2	20	50	0	0	19	51	0	0
L3	0	6	1	0	0	4	3	0
L4	0	0	3	0	0	0	3	0

Table 6.6. Exact GN prediction accuracy and failures by levels

NO$_2$ Level	GN1(FEN)_8				GN2(UGN)_8				GN3(SGN)4_8_5			
	L1	L2	L3	L4	L1	L2	L3	L4	L1	L2	L3	L4
L1	150	20	0	0	150	20	0	0	149	21	0	0
L2	14	52	4	0	27	43	0	0	12	53	5	0
L3	0	3	4	0	0	4	3	0	0	3	4	0
L4	0	0	3	0	0	1	2	0	0	0	3	0

accurately higher levels, in accordance with more adapted and verified explicit rules embedded into the EKM part of the proposed system. That means that in a larger data set, there would be more representative episode cases, closer to the "normal" behavior of the real population cases.

This experiment studied the daily NO$_2$ maximum concentration prediction for the same day, for a single representative measuring station. It would be interesting to study the prediction of the NO$_2$ peak concentration of the next day (as a second requirement from the PERPA experts), using a competitive gating network-based system and representing only the tendencies comparing with the present day.

This proposal is based on the observation made during the tests that such architectures well approximate the tendencies of the output, more than the correct values. Additionally, as a projection of this work, the SO$_2$ and O$_3$ maximum concentration predictions would be a good next testing step. Finally, a global system using close-related prediction modules for all involved measuring stations of the PERPA network is desired in order to extend the prediction task on the spatial axis, based on the already described temporal approach.

6.4.6 A Distributed AI Architecture for Air Quality Monitoring

A good example (Neagu et al. 2001b) of homogeneous integration of our approach for HIS, based on Implicit and Explicit Knowledge Modules, is DNEMO (Distributed NEMO): a framework for supporting problem solving in environmental applications, built over the years by researchers of the Human-Computer Interaction Group of the University of Patras (Kalapanidas and Avouris, 2000). DNEMO is based on a distributed multi-agent architecture and can be connected to an environmental monitoring and prediction network.

The framework has been demonstrated in the field of urban air pollution monitoring, using data from Athens, Greece. The general architecture of the framework is shown in Fig. 6.29. In this figure, the two main clusters of heterogeneous communicating agents of the DNEMO architecture are shown.

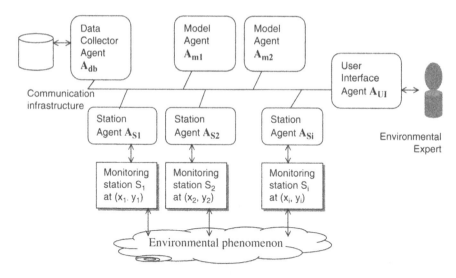

Fig. 6.29. Architecture of the DNEMO environmental problem-solving framework

The *Stations cluster* represents all the spatially distributed monitoring stations. The station agents are homogeneous communicating agents. They are assigned with the task of solving the partial problem of predicting the local air pollution levels. These agents store their own sensory data locally. They contain similar planning methods to resolve data integrity issues and they engage the same strategy to solve their local problem, which represents a partial solution to the overall problem. Each Station Agent maintains a list of other Station and Model Agents capabilities, used for matching in an optimal way a given planned action that cannot be executed locally, with the agent that offers the specified service.

Communication between agents belonging to the same cluster is a crucial part of the overall solution-building phase.

The *Models cluster* contains agents that perform the same tasks in a different way, without communicating with each other. They have the same goal of predicting pollution levels, applying different approaches, as described in the next section of the paper. While for the number of station agents there is a one to one mapping to the physical Monitoring Stations, which are dispersed in various locations of the monitored area, the Model Agents maintain a many to one relationship with the different ML algorithms and knowledge representations that are implemented. A discussion on the performance of some of these agents is included in the following sections of the paper. Apart from these two clusters of agents, three special functionality agents have been built:

- The *TimeWatcher agent* whose duty is to maintain an agenda of scheduled events. For instance it informs relevant agents to initiate a prediction task related to the concentration of a specific air polluting substance.
- The *DataFeeder agent* that is the source of raw data of DNEMO, acting as a facilitator between DNEMO agents and the database of the Air Quality Management Center.
- The *User Interface agent*, which communicates with the end-user. It is there where all predictions from all station agents are gathered and presented to the human operator.

The basic architecture of the agents facilitating their collaborative behavior and their communications is presented next. The LALO agent oriented platform (Gauvin 1995) has been used for building the DNEMO framework. The characteristics of this environment, are common to many similar available tools. LALO is an object-oriented environment that contains an extendable set of classes, from which application agents can be built. However even these simple agents have a behavior and communication component, as shown in Fig. 6.30.

The later through threads, permit the asynchronous processing of incoming and outgoing message queues by the agent. The messages sent and received by an agent are based on the KQML protocol (Finin et al. 1998). The simple reactive agent can process messages of *transport-address* and *achieve*

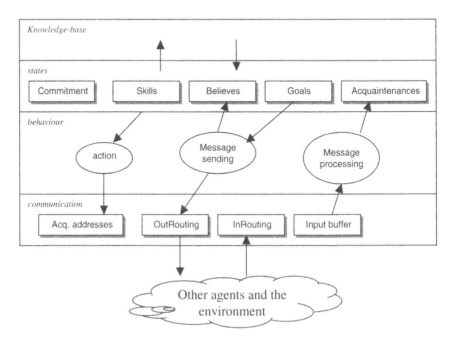

Fig. 6.30. Agent architecture

types only. A more complex agent contains also a mental component. This component permits representation of an agent's belief in its skills (that is the tasks that the agent can undertake, divided into private tasks and public ones which can be executed after request for other agents), its goals and commitments.

The agent also maintains models of other agents (acquaintances), as shown in Fig. 6.30. This agent can have a more complex behavior than that of the reactive agent. For instance through the commitments structure, the agent can commit itself to a plan and therefore reject a request for a new job made by another agent. This agent can respond to messages of type *stream-about, ask-if, tell* etc. Even more complex agents can be built that contain a control mechanism of their behavior, represented in Fig. 6.30 as *Agent Knowledge Base*. This can contain rules controlling the processing of the incoming messages, the mental state of the agent and the control of the agent's actions. The LALO agents can be programmed through the LALO programming language, in which the beliefs in the private and public tasks and the other characteristics of the agent can be defined.

6.4.7 Conclusions

The proposed neuro-fuzzy knowledge-based model represents an encouraging alternative to the stochastic models. We proved these kinds of models are able to learn sophisticated air pollution dispersion mechanisms from basic meteorological and air pollution observations. In addition, they are capable of representing knowledge acquired from human experts in order to improve the prediction results for situations not used in the training process. By interpreting the results of the conducted experiments, we can conclude that the proposed structures and specific methods compete with each other very close, and they constitute an efficient decision support system that could function as a predictive tool in an air quality operational center. The results of the proposed versions of the neurosymbolic-integrated system are both, qualitative, and quantitative, and they can indicate a reasonable decision, and specific motivations. Our experiments show that, in complex and multidimensional temporal and spatial factors-described domains of real world problems, as it is the case of air quality prediction, neuro-fuzzy integration represents a better alternative to classical artificial intelligence tools, mainly due to the strong dependencies of input data.

The overall performance of the proposed approach can be described as significant, considering that the human experts at the Greek Air Quality Operational Center do not exceed these resulted accuracy levels. This fact makes the proposed system and its versions useful for operational conditions, as well as adaptable for other similar classification or prediction problems. The adaptive nature of the basic structure, as well as the additional properties to extract acquired knowledge, which was proven even in the case of the relatively limited data set used, are also important features of the proposed hybrid architectures. Moreover, the homogeneity imposed by the MAPI unit-based implementation, exhibit a prospective approach of hybrid systems. The results reported here provide an indication that neurosymbolic integration can be a very useful tool for researchers and practitioners in air quality management and prediction.

All the proposed methods and structures are easily implemented; they deliver prediction at run time and do not require a dense monitoring network. Based on the fuzzy computation performed by the MAPI neural unit, all the proposed versions of the neurosymbolic system can deal with input noise and uncertainty, and do not require expensive equipment. On the other hand, the proposed hybrid system, and the specific architectures combining implicit and explicit knowledge, contribute to the understanding of the phenomenon, and produce qualitative as well as quantitative indications. The proposed structures permit knowledge acquisition and knowledge extraction, which means that the initial architecture can be improved and adapted. The encouraging results of application of the described models in air quality prediction indicate that the method is worth further research and extension to other similar real life problems with similar characteristics.

6.5 A Modular Intelligent Knowledge-Based System for Predictive Toxicology: NIKE

We are increasingly confronted with the consequences of the action of toxic chemical compounds on human health and on the environment. A number of studies have been based upon the premise that toxicity is related to the physico-chemical properties of a compound. Recent advances in the field of Artificial Intelligence, particularly Hybrid Intelligent Systems, have provided new representation, modeling and processing possibilities to this area of research.

6.5.1 Introduction

We are becoming increasingly aware of the need to understand and predict the consequences of waste chemicals to human health and environment. The aim of predictive toxicology is to estimate if a chemical compound can have harmful effects on a biological target. This is presently done through experiments which are expensive in terms of financial cost and time, and involve animal studies using in vivo procedures. The number of substances currently listed by the Chemical Abstracts Service (www.cas.org) exceeds 23 millions, a figure continuously increasing. Because of the ever expanding use of chemicals within productive processes, a better understanding of their toxicological impact on human life, plant life, wild-life, and the environment in general, is mandatory. Methodical and thorough experimental evaluation of the toxicity of chemicals requires excessive time and huge economical resources. In addition to critical ethical considerations associated with the experimental use of animals, characterizing a single substance by these methods costs millions of pounds and takes many years. The huge number of chemical compounds, including the case of the waste compounds, to be studied makes this especially challenging. In many cases a single chemical compound can generate many transformation products that are released in the environment, or are exposed to animals or man over many years. Each of these transformation products requires, in principle, the same attention devoted to the parent compound. As a consequence, the number of classes of compounds to be studied becomes enormous. Thus, alternative procedures have to be considered.

The problem of toxicity prediction represents both the application of our approach and the inspiration of some solutions here reviewed. Toxicology is the science that defines the limits of safety of chemical agents. Predictive toxicology is an excitingly novel area. Many researchers (Schultz et al. 2003) investigated new molecular properties and descriptors, trying to find better relationships. A number of studies have been based upon the premise that toxicity is related to physico-chemical properties of a compound (Cronin and Schultz 2003): so far, up to 3,000 chemical descriptors have been investigated. Most of these are global descriptors, i.e. describe the total compound. In some cases, local descriptors are used, which refer to some part of the molecule, such as a fragment. This high number of descriptors represents a problem and several methods have been proposed to select and use descriptors, including linear (such as principal component analysis) or non-linear systems (such as machine learning algorithms). Recent advances in the field of describing molecular properties and in artificial intelligence (AI), particularly hybrid intelligent systems (HIS), have provided possibilities to improve this area of research.

6.5.2 Predictive Toxicology

Predictive toxicology is a multi-disciplinary science that requires close collaboration among toxicologists, chemists, biologists, statisticians and AI/ machine learning researchers (Neagu and Gini 2003). The goal is to describe the relationship between the chemical structure and biological and toxicological processes. This relationship is known as the structure – activity relationship (SAR).

Following the conclusion in the late 1990s that toxicity was an important cause of costly late-stage failures in drug development, it has become widely accepted that toxicological evaluation should be considered as early as possible in the environmental risk assessment of new chemical compounds and also in the drug discovery process. In toxicity prediction there are many variables: the toxicological endpoint, the number of molecules in the data set, the homogeneity of the data set, the methods to describe the physico-chemical properties of molecules, the computational algorithms to produce statistical relationships, and the validation method. An endpoint refers to a specific organism, with a specific route of administration (e.g. oral) and a specific exposure time.

A quantitative structure – activity relationship (QSAR) is a mathematical model that relates the quantitative measures of a chemical structure (e.g. a physicochemical property) to a biological effect (e.g. toxicological endpoint). These relationships can be developed mostly through a regression process (Schultz et al. 2003; Cronin and Schultz 2003; Hansch et al. 1995). From theory and experience, one could define the postulates of modeling QSARs (Craciun and Neagu 2003):

P1: The molecular structure is responsible for all the activities.
P2: Similar compounds have similar biological and chemo-physical properties.
P3: QSAR is applicable only to similar compounds.

The traditional way of assessing the toxic risk of a compound is to test it on animals. The results are then extended to humans using safety factors and dose relationships. This approach, however, suffers many drawbacks: cost of the experiments (>1 million US$ per compound), the duration of the tests (3–5 years), public pressure to reduce or eliminate the use of animals in scientific experiments.

Why Hybrid Intelligent Systems in Predictive Toxicology?

Various AI approaches were proposed in the last few years to be applied to toxicity prediction. Some authors have focused on the use of neural networks to predict toxicity of chemical compounds due to their capabilities to cope with complexity, uncertainty, noisy or corrupted data (Gini 2000; Craciun and Neagu 2003). Two QSAR series were analyzed using neural networks; the results were compared with other three AI-related approaches (Adamczak and Duch 2000). A hybrid expert system approach was done by (Gini et al. 2001) and applied to predict phytotoxicity. A study on the usage of fuzzy logic for descriptors modeling has been presented by (Exner and Brickmann 1997). In all cases the neural network approach of the toxicity prediction is restricted to crisp modeling of data. There is also a tremendous number of authors who tried to integrate neural networks and fuzzy logic with application to various other fields (Buckley and Hayashi 1995; Hatzilygeroudis and Prenzas 2001; Neagu et al. 2001), but very few tried to apply combined neuro-fuzzy techniques to predictive toxicology problems.

Neural networks are very good modeling tools for highly non-linear and complex data and they can be seen as a universal approximation technique. Fuzzy techniques can be appropriate for analyzing toxicity data as it allows the integration in a natural way of the toxicology knowledge into the process of data interpretation and analyzing. The formulation of the knowledge is done in a human understandable way such as linguistic rules. The main drawback of neural networks is represented by their "black box" nature and lack of transparency in human understandable terms, whilst the disadvantage of fuzzy systems is represented by the difficult and time-consuming process of knowledge acquisition. On the other hand the advantage of neural network over fuzzy systems is learning and adaptation capabilities, while the advantage of fuzzy system is the human understandable form of knowledge representation. Neural networks use an implicit way of knowledge representation whilst fuzzy and neuro-fuzzy systems represent knowledge in an explicit form, such as rules.

QSARs as Fuzzy Inference Rule-based Systems

Finding a QSAR (Quantitative Structure-Activity Relationship) is essentially a regression process and, historically, linear regression methods have been used (Schultz et al. 2003). Some regression-based models results in exhibiting instability when trained with noisy data.

Following mapping procedures to represent fuzzy inference rules as connectionist structures (Neagu et al. 2002a), a number of explicit knowledge modules implementing first-order Sugeno fuzzy models (Takagi and Sugeno 1985) will describe subdomains of a given QSAR. The number of EKMs will be equal with all the combinations of fuzzy values of inputs and outputs represented in the specific QSAR. The output of this collection of EKMs will be a single MAPI neuron, acting as an arithmetical device (da Rocha 1992). This mechanism permits the implementation of neuro-fuzzy modules, equivalent with QSARs.

6.5.3 Applications of Hybrid Intelligent Systems in Predictive Toxicology

There are few data sets used to build up case studies based on the Hybrid Intelligent Systems structures described in the previous sections:

- The U.S. Environmental Protection Agency provided a data set, starting from a revision of experimental data from literature, and which referred to acute toxicity 96 hours (LC50), for fathead minnow (Pimephales promelas). The target is to predict the toxicity as $\log(1/LC50)$ based on a collection of chosed descriptors from the original data set of 150 descriptors (Neagu et al. 2002a).
- An analysis of 225 phenolic toxicity data to *Tetrahymena pyriformis* (Neagu et al. 2002b). The most successful architectures of the developed neural and neuro-fuzzy models were applied for classes of mechanism of action of phenols, in order to obtain specific models and to compare with the traditional QSAR approach. A study to identify the most useful models is also described.

– A model based on neuro-fuzzy networks is developed to provide an improvement in combining the results of five classifiers applied in toxicity of pesticides (Benfenati et al. 2002).

Modeling Acute Aquatic Toxicity

The U.S. Environmental Protection Agency provided to build up a data set, starting from a revision of experimental data from literature, referred to acute toxicity 96 hours (LC_{50}), for fathead minnow (*Pimephales promelas*). An accurate analysis of the experimental information will permit us to associate a mode of toxic action (MOA) to each compound. The data set contains 568 organic compounds common to in industrial processes. It is a large set of compounds belonging to different chemical classes: a positive characteristic is the homogeneity and reliability of this toxicological data.

A set of about 150 descriptors was examined (Neagu et al. 2002). The descriptors are classified, according to CODESSA (Katritzky et al. 1994) into constitutional descriptors (34), depending on the number and type of atoms, bonds and functional groups; geometrical descriptors (14), which give molecular surface area and volume, moments of inertia, shadow area, projections and gravitational indices; topological descriptors (38), which are molecular connectivity indices, related to the degree of branching in the compounds; electrostatic descriptors (57), such as partial atomic charges and others depending on the possibility for some sites in the molecule to form hydrogen bonds; quantum-chemicals descriptors (6), i.e. total energy of the molecule, the energies of the lowest unoccupied and highest occupied orbital (HOMO and LUMO), ionisation potentials, heat of formation; hy-drophobic descriptors (7), which are logP, logD, the expression of lipophilicity of the molecule at various pH.

For fuzzy processing, the membership functions were selected to simplify the calculus and to reduce the number of input neurons: all the descriptors followed a trapezoidal fuzzification. The linguistic variables considered for toxicity (Fig. 6.31) are characterized by the following term sets:

The descriptors selection was obtained by Principal Components Analysis (PCA) and Correlation Analysis techniques (Table 6.7):

For the current data set, three original QSAR equations are used (Neagu et al. 2002a). Figure 6.32 depicts the behaviour of the QSAR2 as Sugeno-type fuzzy inference system (predicted versus observed values for LC_{50}) with the following three generated rules:

1. If (*logP* is Low) then (*log1/LC50* is QSAR2) (AND)
2. If (*logP* is Med) then (*log1/LC50* is QSAR2) (AND)
3. If (*logP* is High) then (*log1/LC50* is QSAR2)

The modules used to acquire knowledge about toxicity related to the 17 chosen descriptors are implicit knowledge-based IKMs. The *implicit knowledge* modules are represented by the selected ANN/FNN, created and adapted

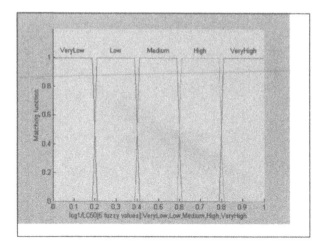

Fig. 6.31. Linguistic values of toxicity

Table 6.7. Preselected descriptors

Descriptors	Code
Total Energy (kcal/mol)	QM1
Heat of Formation (kcal/mol)	QM3
LUMO (eV)	QM6
Relative number of N atoms	C9
Relative number of single bonds	C24
Molecular weight	C35
Kier&Hall index (order 0)	T6
Average Information content (order 1)	T22
Moment of inertia B	G2
Molecular volume	G10
Molecular surface area	G12
TMSA Total molecular surface area	E13
FPSA-2 Fractional PPSA (PPSA-2/TMSA)	E24
PPSA-3 Atomic charge weighted PPSA	E28
FPSA-3 Fractional PPSA (PPSA-3/TMSA)	E31
logD	pH9
logP	logP

by learning algorithms. The representation of implicit knowledge is based on the numerical weights of the connections between neurons.

The first module, called IKM-CNN (Implicit Knowledge Module-based on Crisp Neural Networks), takes charge of modelling the data set as a multilayer perceptron (MLP). The MLP model is used to compare the overall performance of the neurosymbolic system with neuro-fuzzy and QSAR approaches.

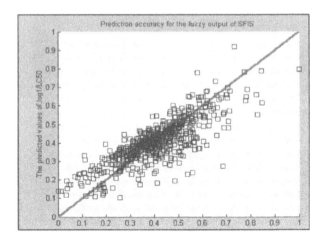

Fig. 6.32. Predicted versus observed values: Sugeno FIS in neuro-fuzzy representation, equivalent to QSAR2

The second module, IKM-FNN (Implicit Knowledge Module-based on Fuzzy Neural Networks) is implemented as a multilayered neural structure with an input layer, establishing the inputs to perform the membership degrees of the current values, a fully connected three-layered fuzzy neural network, and a defuzzification layer, as discussed in Chap. 3. The weights of the connections between layer 1 and layer 2 are set to one.

An innovative and immediate result is the application of the two developed models in the steps 1–3 (trained FNN25HN and CNN35HN) for studying the descriptors significance (Fig. 6.33). Understanding the relevance of descriptors in the context of toxicity prediction is a tedious task, considering the big number of descriptors and compounds to be studied. We replaced, in the generated model, the studied descriptors column with zero (Gori and Scarselli 1998) in the test data set (Fig. 6.33).

The results depict that IKM-FNN models are more robust to noisy data than CNNs: they are more suitable to toxicity classification and if a specific method of defuzzification is used (generally, the mean of maximum method replaces the centroid of area method).

The IKM-CNN models are more sensitive to the noisy data, as described, which make them an important indicator of the significance of the descriptors to toxicity. CNN shows two behaviours about missing descriptor, other than a normal small increasing of absolute prediction error (Fig. 6.33a,c,e): predictions translation (Fig. 6.33b,f), announcing linear dependence with the absent descriptor, or a proportional magnify of error, consequence of a nonlinear relation of some of the current inputs (predictions rotation, as presented in Fig. 6.33d).

The final structures of the model are based on the three strategies to build a HIS, described in Chap. 5: FEM (fire each module using statistical,

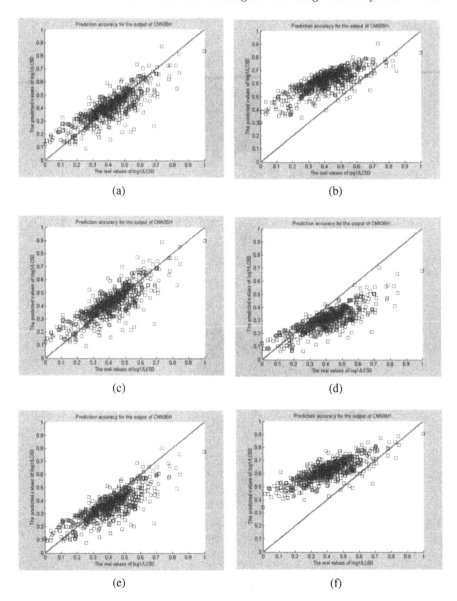

Fig. 6.33. Performance validation (predicted data set versus real data values) for: (a) complete test data set; not significant descriptor missing in test data set $QM1$ (c), or $C9$ (e); significant descriptor missing in test data set $T6$ (b), $G2$ (d) or the most important $logP$ (f)

FEMS, and fuzzy, FEMF, integration, as discussed in Neagu and Gini 2003) having the results presented in Fig. 6.34, UGN (unsupervised-trained gating network for all the implied modules' fusion) and SGN (supervised-trained

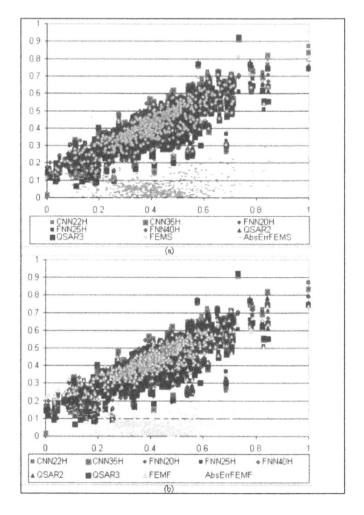

Fig. 6.34. Observed versus predicted values for the individual experts and the combination, including the absolute error distribution: (**a**) FEMS; (**b**) FEMF

gating network to integrate the expert modules), with results presented in Fig. 6.35.

The presented evaluation shows that neural/neuro-fuzzy networks can learn patterns of organic compounds. Further, hybrid structures can be used to predict the behaviour of such chemicals, to classify under a toxicity scale and to select the most useful descriptors. The evaluation shows that these predictions are up to 5% more accurate than those of the classical approaches. This offers the possibility to improve the performance by using neuro-fuzzy ensembles to guide and refine further decisions.

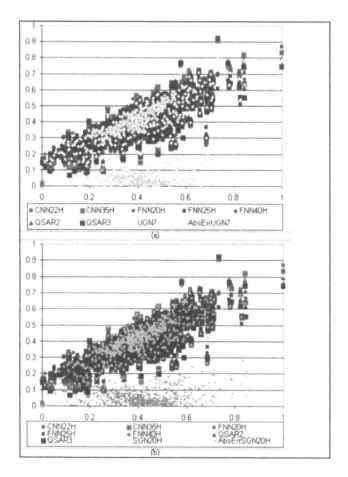

Fig. 6.35. Observed versus predicted values for the individual experts and the combination, including the absolute error distribution: (a) UGN; (b) SGN

Significant improvement was found for the use of neural ensembles, since explicit knowledge was inserted in the system. This suggests that the 568 compounds used in this study do not provide a best coverage of the problem domain, which is split in contiguous sub domains. A good model, that does not overfit to training data, should be built with more than one ANN/FNN.

Modeling Phenolic Toxicity

Biological Data

The 2D ciliate (*Tetrahymena pyriformis*) population growth impairment (IGC50) data were processed (Neagu et al. 2002b) from the TETRATOX database described by Schultz (Schultz 1996). All chemicals and their up-dated toxicity values are available as IMAGETOX (Gini et al. 2002) data sets. Toxicity was quantified as the log of the inverse of potency. Potency values (reported as millimoles, mM) were normalized to their extremes of the set.

Molecular Descriptors

A total of 158 2D and 3D descriptors were calculated for each compound. Hydro-phobicity with and without correction for ionization ($\log P$, $\log D$), acidity constant (pKa) and energy of the lowest unoccupied molecular orbital (E_{LUMO}) are the most cited entry variables for the problem of toxicity prediction in our case. Some of them were rejected, due to the zero values for most of the chemicals. Finally, 43 descriptors were used as inputs in ANNs, and the output was toxicity as $\log(1/IGC_{50})$. From these, a number of 7 descriptors built the second set of entries, to study their significance in a final reduced size model of toxicity and correlation with MOA.

The mode of action (MOA) of the phenols is coded in the following classes: MOA $= 1$, (153 polar narcotics: 4-hydroxyphenylaceticacid, 1,3,5-trihydroxybenzene, etc.), MOA $= 2$ (18 respiratory uncouplers: 2,4,6-trinitrophenol, 2,4-dichloro-6-nitrophenol, etc.), MOA $= 3$ (27 pro-electrophiles: 2,4-diaminophenol2HCl, 3-methylcatechol, etc.), MOA $= 4$ (23 soft electrophiles: 3-nitrophenol, 4-hydroxy-3-nitrobenzaldehyde, etc.) MOA $= 5$ (4 pro-redox cyclers: tetrabromocatechol or tetrafluorohydroquinone).

A two-parameter QSAR, or response-surface, was developed by Cronin and Schultz (1996) based on parameters for hydrophobicity and electrophilicity for the toxicity of a limited selection of these compounds (QSAR1):

$$\log(IGC_{50})^{-1} = 0.67(0.02)\log P - 0.67(0.06)LUMO - 1.123$$
$$n = 120,\ R^2 = 0.90,\ R^2_{CV} = 0.89,\ s = 0.26,\ F = 523 \quad (6.2)$$

where IGC_{50} is the concentration in millimoles causing 50% inhibition of growth, after 40 hours, to *Tetrahymena pyrformis*, P is the octanol-water partition coefficient, $LUMO$ is the energy of the lowest unoccupied molecular orbital, n is the number of observations, R^2 is the coefficient of determination, R^2_{CV} is the leave-one-out cross-validated coefficient of determination, s is the standard error of the estimate and F is the Fisher statistic. Figure in parentheses are the standard errors on the coefficients.

Garg, Kurup, and Hansch (2001) obtained, on the same dataset, a similar relationship (QSAR2), replacing $LUMO$ with Hammett constant (σ):

$$\log(IGC_{50})^{-1} = 0.64(0.04)\log P - 0.61(0.12)\sigma + 1.123(0.13) \quad (6.3)$$
$$n = 119,\ R^2 = 0.90,\ R^2_{CV} = 0.89,\ s = 0.265$$

Cronin et al. 1992, working with a greater number of phenols to *Tetrahymena pyriformis*, demonstrated that the $\log D$ and $LUMO$ are the most successful descriptors in modeling (QSAR3) of phenols toxicity (correlation coefficient between $\log D$ and $LUMO$ is 0.396):

$$\log(IGC_{50})^{-1} = 0.60(0.021)\log D - 0.69(0.058)LUMO$$
$$-0.81(0.054) \tag{6.4}$$
$$n = 161, R^2 = 0.84, R^2_{CV} = 0.83, s = 0.33, F = 420$$
$$T\text{-}values : \log D = 28.3, LUMO = 12.0$$

Data Preparation and Analysis

The whole set of available patterns was divided in two independent sets, each one of them having its own task in the model training and testing processes (Fig. 6.36). A pattern is defined as a vector of values of the input features (selected descriptors) and values of the output, toxicity of phenols.

The training set was used for the adjustment of the connections of the neural and neuro-fuzzy networks with a backpropagation (*traingdx*) algorithm; *traingdx* is a network training function that updates weight and bias values according to gradient descent momentum and an adaptive learning rate.

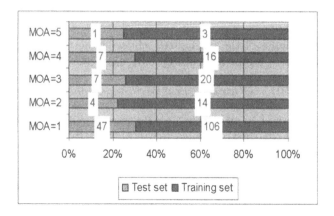

Fig. 6.36. The distribution of training/testing sets against the MOA classes

The proposed hybrid system contains modules that could be used individually, and operate on the same inputs in order to model and to predict the phenols' toxicity. Both the input data set values and the output ones, are normalized and fuzzified with respect to the 225 phenols descriptors values. The input data set consists of 43 descriptors, while the output is toxicity: $\log(1/IGC_{50})$.

For FNN processing, the membership functions were considered to simplify the calculus and to reduce the number of involved input neurons. All the descriptors followed a fuzzification trapezoidal-triangular-trapezoidal (Fig. 6.37). Consequently, the linguistic variables considered for descriptors inputs are characterized by the term sets:

$$D_i = \{Low, Med, High\}, i = 1.43 \qquad (6.5)$$

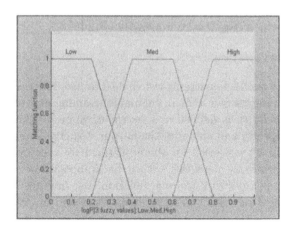

Fig. 6.37. Fuzzy shapes used to describe descriptors as linguistic values

The fuzzy shapes of the normalized values of the toxicity considered as linguistic variable are presented in Fig. 6.31, while the terms set is:

$$\log(1/IGC50) = \{VeryLow, Low, Medium, High, VeryHigh\} \qquad (6.6)$$

Five levels of toxicity are defined for the normalized $\log(1/IGC_{50})$: *Very-Low* (0–0.2), *Low* (0.2–0.4), *Medium* (0.4–0.6), *High* (0.6–0.8), and *VeryHigh* (0.8–1). The slopes of the shapes were considered in order to interpret the outputs as classification or as continuous values. The membership functions shapes range is: *Bell, Gaussian, Pi, S, Z, Triangular, Trapezoidal,* and *Sigmoidal.*

The whole set of available patterns was divided in two independent sets, each one having its own task in the model training and testing processes (Fig. 6.38). A pattern is defined as a vector of values of the input features (selected descriptors) and values of the output, toxicity of phenols.

The training set was used for the adjustment of the connections of the neural and neuro-fuzzy networks with a backpropagation (traingdx) algorithm. The testing set was used for testing both the trained neural and neuro-fuzzy networks. In order to determine the performance of the overall best model, the same testing set was used as a production set of data.

The data set (225 compounds) was divided in such a way to conserve the distribution of the 5 classes of MOA, as well as the five fuzzy values of the output linguistic variable (Fig. 6.31). The algorithm was a 70–30 partitioning, as it is used in the majority of such kind of comparative tests between predictive algorithms: 159 training cases and 66 testing cases.

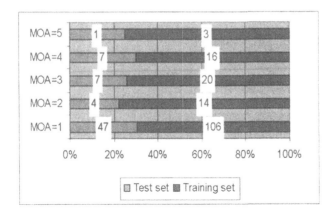

Fig. 6.38. The distribution of training/testing sets against the MOA classes

The Integrated Models

The designed neuro-fuzzy networks are multi-layered structures with the 43×3 above described fuzzy inputs, 5 fuzzy output neurons and the toxicity linguistic variable $\log(1/IGC_{50})$. When the number of hidden neurons parameterized the neural/neuro-fuzzy networks a medium number is desirable.

For an ANN to be able to generate closed decision regions, the minimum number of hidden units must be greater than the number of input units (Gori and Scarselli 1998). To derive the maximum number of hidden units in the network, results based on Kolmogorov's theorem were used. (Hecht-Neilson 1990) established that the maximum number of hidden neurons needed to represent any function of n variables is less than twice the number of inputs $2 \times n\_input + 1$.

Backpropagation algorithm was used for training (Rumelhart and Mc-Clelland 1986). A learning rate of 0.7 and a momentum term of 0.9 were used (a relatively high learning rate ensures rapid finding of the error function minimum and a high momentum term prevents too many oscillations of the error function). Networks were trained up to 5000 epochs, giving an error about 0.005. Similar context was applied for CNN training. Finally, the 90 hidden neurons IKM-CNN (CNN90H) and 20 hidden neurons IKM-FNN (FNN20H) were considered (Fig. 6.39).

Results on Phenolic Toxicity Modeling with Hybrid Intelligent Systems

The performances of the models were evaluated on the testing data set. The outputs of the explicit and implicit modules, viewed as inference results, are computed for a given testing pattern. A typical model performance is shown in Fig. 6.40 (the predicted toxicity value: 0.74745, the real toxicity: 0.75902).

Consequently, the connectionist approaches were used as descriptors correlation test tool. The descriptors for the reduced data set are: CX_EPM20,

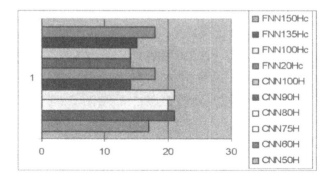

Fig. 6.39. The number of outliers with an absolute error prediction greater than 0.15, for various IKM-CNN and IKM-FNN architectures

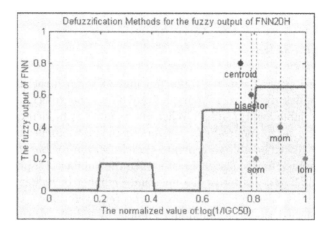

Fig. 6.40. The prediction of toxicity as a fuzzy inference on IKM-FNN

MO_Dmax, MO_Amax, TS_LUMO, TS_HOMO, QS_SHHBd, QS_SHBa. Good results were obtained (Table 6.8). The accuracy values are related to the absolute errors as follows: Accuracy1 is relative to absolute error>0.15, Accuracy2: to absolute error>0.2, Accuracy3: to absolute error>0.3. In the case of the reduced number of inputs, all the connectionist networks were more difficult to train and the number of outliers was bigger.

Table 6.8. The distribution of testing and training sets

	QSAR2	QSAR3	QSAR4	CNN90H	FNN20H
Accuracy1	0.5622	0.5660	0.5471	0.9377	0.9377
Accuracy2	0.7283	0.7433	0.7358	0.9822	0.9511
Accuracy3	0.9471	0.9433	0.9433	0.9955	1.0000

Consequently, classification of the toxicity correlated to MOA for phenols requires a high degree of experience from computational chemistry experts. The several described approaches have generated suitable computer-based classifiers for these patterns. The described classifiers range from a QSAR to a neuro-fuzzy system, through TO classical ANN architectures (Table 6.8). The main problem regarding the symbolic approach is the difficulty of improvement and correlation analysis, due to the existence of limitations in knowledge elicitation, as this is a complex domain. Several implicit knowledge models with different number of neurons on the hidden layer were trained, analyzed and added to a final intelligent hybrid model for better results.

Analysis of Classifiers of Pesticides Toxicity Using Hybrid Intelligent Systems

Classification systems for QSAR studies are quite usual for carcinogenicity (Gini et al. 2001), because carcinogenicity classes are defined by regulatory bodies such as IARC and EPA. For ecotoxicity, most of the QSAR models are regressions, referring to the dose giving the toxic effect in 50% of the animals (for instance LC_{50}: lethal concentration for 50% of the test animals). This dose is a continuous value and regression seems the most appropriate algorithm. However, classification affords some advantages. Indeed, (i) the regulatory values are indicated as toxicity classes and (ii) classification can allow a better management of noisy data. For this reason the subject of investigation is classification. No general rule exists to define an approach suitable to solve a specific classification problem. In several cases, a selection of descriptors is the only essential condition to develop a general system. The next step consists in defining the best computational method to develop robust structure-activity models.

Data Preparation and Models

A data set consisting of 57 common organophosphorous compounds has been investigated (Benfenati et al. 2002). The main objective is to propose a good bench-mark for the classification studies developed in this area. The toxicity values are the result of a wide bibliographic research mainly from the *"Pesticide Manual"*, and the ECOTOX database system. An important problem was connected with the variability that the toxicity data presents. Indeed, it is possible to find different values showing for the same compound and the same end-point LC_{50} different for about two orders of magnitude. Such variability is due to different factors, as the different individual reactions of organisms tested, the different laboratory procedures, or is due to different experimental conditions or accidental errors.

The toxicity value was expressed as $Log_{10}(1/LC_{50})$. Then the values are scaled in the interval $[-1.1]$. Four classes are defined: Class 1 $[-1 \ldots -0.5)$,

Class 2 $[-0.5\ldots0)$, Class 3 $[0\ldots0.5)$, Class 4 $[0.5\ldots1]$. A set of 150 descriptors werecalculated. They were split into six categories: Constitutional (34), Geometrical (14), Topological (38), Electrostatic (57), Quantum-chemicals (6), and Physico-chemicals (4). A selection of the variables, to better describe the molecules, is necessary to obtain a good model. There is the risk that some descriptors do not add information, increase the noise, and make more complex the result analysis. Furthermore, using a relatively low number of variables, the risk of overfitting is reduced. The descriptors selection was obtained by Principal Components Analysis (PCA).

The classification algorithms used for this work are five: LDA (Linear Discriminant Analysis), RDA (Regularized Discriminant Analysis), SIMCA (Soft Independent Modeling of Class Analogy), KNN (K Nearest Neighbors classification), CART (Classification And Regression Tree). The first two are parametric statistical systems based on the Fisher's discriminant analysis, the third and fourth are not parametrical statistical methods; the last one is a classification tree.

LDA: the Fischer's linear discrimination is an empirical method based on p-dimensional vectors of attributes. Thus the separation between classes occurs by a hyperplane, which divides the p-dimensional space of attributes.

RDA: The variations introduced in this model have the aim to obviate the principal problems that afflict both the linear and quadratic discrimination. A more efficient regulation was carried out by Friedman, who proposed a compromise between the two previous techniques using a biparametrical method for the estimation (λ and γ).

SIMCA: the model is one of the first used in chemometry for modeling classes and, contrarily to the techniques before described, is not parametrical. The idea is to consider separately each class and to look for a representation using the principal components. An object is assigned to a class on the basis of the residual distance, rsd^2, that it has from the model which represent the class itself:

$$r_{ijg}^2 = (\hat{x}_{ijg} - x_{ijg})^2, rsd_{ig}^2 = \left(\sum_j r_{ijg}^2\right) \bigg/ (p - M_g) \qquad (6.7)$$

where $\hat{x}_{igj}$ = co-ordinates of the object's projections on the inner space of the mathematical model for the class, x_{igj} = object's co-ordinates, p = number of variables, M_g = number of the principal components significant for the g class, $i = 1, \ldots, n$ = number of objects, $j = 1, \ldots, p$ = number of variables.

KNN: this technique classifies each record in a data set based on a combination of the classes of the k record(s) most similar to it in a historical data set ($k = 1$).

CART is a tree-shaped structure that represents sets of decisions. These decisions generate rules for the classification of a data set. CART provides a set of rules that can be applied to new (unclassified) data set to predict which records will have a given outcome. It segments a data set by creating two-way splits.

The more common methods for validation in our approach are: (i) Leave-one-out (LOO); (ii) Leave-more-out (LMO); (iii) Train & Test; (iv) Bootstrap. We used LOO, since it is considered the best working on small dimension data sets (Helma et al. 2000). According to LOO, given n objects, n models are computed. For each model, the training set consists of $n - 1$ objects and the evaluation set consists of the object left. The Non Error Rate percentage (NER%, Table 6.9) represents the ability of the algorithm in modeling the problem (fitting) and predicting his value (validation, using LOO). To estimate the predictive ability, we considered the gap between the experimental (fitting) and predicted value (cross-validation) for the n objects left, one by one, out from the model.

Table 6.9. Performances of the classification algorithms

	NER% Fitting	NER% Validation	Descriptors
LDA	64.91	61.40	D1, D2, D3, D4
RDA	84.21	71.93	D1, D2, D3, D4, D6, D7, D8, D11, D12, D13
SIMCA	92.98	77.19	D1, D2, D3, D4, D5, D6, D7, D8, D10, D11, D12
KNN	–	61.40	D1, D12
CART	85.96	77.19	D1, D2, D3, D4, D5, D9

The Hybrid Intelligent System Combining Classifiers Prediction

Combining multiple classifiers could be considered as a direction for the development of highly reliable pattern recognition systems, related to the hybrid intelligent systems (HIS) approach. Combination of several classifiers may result in improved performances (Duin et al. 2000). The necessity of combining multiple classifiers is arising from the main demand of increasing the quality and reliability of the final models. There are different classification algorithms in almost all the current pattern recognition application areas, each one having certain degrees of success, but none of them being as good as expected in applications. In this section we will discuss a combination technique for the toxicity classification as a neuro-fuzzy gating the output of implied classifiers, trained against real classification values (Benfenati et al. 2002). This approach allows multiple classifiers to work together.

Every class was represented by the centroid of each of the four classes in which the available domain was split: 0.135 (class 1), 0.375 (class 2), 0.625 (class 3), and 0.875 (class 4). The inputs and the output followed a trapezoidal (de)fuzzification (Fig. 6.16): *VeryLow* $(0 \ldots 0.25)$, *Low* $(0.25 \ldots 0.5)$, *Medium* $(0.5 \ldots 0.75)$, *High* $(0.75 \ldots 1)$.

An Implicit Knowledge Module was implemented, where $p = 5$ inputs represent the classification of the above described algorithms for a given compound: $x_1 = \text{output}_{\text{CART}}$, $x_2 = \text{output}_{\text{LDA}}$, $x_3 = \text{output}_{\text{KNN}}$, $x_4 = \text{output}_{\text{SIMCA}}$, $x_5 = \text{output}_{\text{RDA}}$.

The neuro-fuzzy network has been trained on a set of 40 cases (70% of the entire set). The neuro-fuzzy network is a multilayered structure with the 5×4 above described fuzzy inputs and 4 fuzzy output neurons, the toxicity class linguistic variable (Figs. 6.31 and 6.37). Since different models (5 to 50 hidden units) were built, the best processing modules were found the IKM-FNN with 10, 12, 19 neurons (Fig. 6.42a,b).

A momentum term of 0.95 was used (to prevent too many oscillations of the error function). The nets were trained up to 5000 epochs, giving an error about 0.015. The recognition error for the above models is 5.26% (Tables 6.10, 6.11, Fig. 6.42a,b). Table 6.11 shows the three wrong predictions of the best trained versions of the system, used to calculate the accuracy: two of them are identical, as input values (Chlorpyrifos and Profenofos).

Table 6.10. Confusion matrix for the neuro-fuzzy combination of classifiers

		Assigned Class				Number of Objects
		1	2	3	4	
True Class	1	13	2			15
	2		20			20
	3		1	15		16
	4				6	6

Table 6.11. True and assigned classes for the wrong predictions of the neuro-fuzzy combination of classifiers

	True Class	CART	LDA	KNN	SIMCA	RDA	FNN
Chlorpyrifos	1	2	2	1	2	2	2
Profenofos	1	2	2	1	2	2	2
Fenitrothion	3	2	3	3	3	3	2

The confusion matrix shows the ability to predict in our approach. The best performance was obtained by SIMCA (Table 6.11), which would correctly classify almost 93% of the molecules. This encouraging result was obtained with whole data set involved in developing the model. If we take a

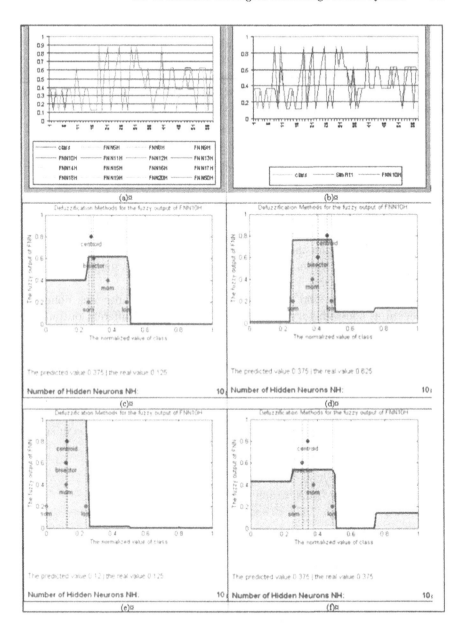

Fig. 6.41. The results of training FNNs: (**a**) 3–5 errors, the best are FNN10H, FNN12H and FNN19H; (**b**) the chosen model, FNN10H, against the SIMCA results and the real ones; (**c**) the bad fuzzy inference prediction for 2 cases in class 1 (Chlorpyrifos and Profenofos); (**d**) the bad fuzzy inference prediction for the case in class 3 (Fenitrothion); two samples of good prediction for test cases: (**e**) a class 1 sample (Phorate); (**f**) a class 2 sample (Edinfenphos)

look to the NER% validated with LOO, we can notice that we loss a lot of the reliability of the model when we predict the toxicity of an external object. Such a behavior proves the ability in modeling of these algorithms, but shows also their incapacity in generalization. The neuro-fuzzy approach seems to overcome this problem, succeeding in voting for the best opinion and having better results than all individual classification algorithms.

Mining the Results

Some relevant fuzzy rules are extracted from the IKM-FNN structures using the Effect Measure Method (Enbutsu et al. 1991; Jagielska et al. 1999). The EMM combines the weights between the layers of the proposed structure in order to select the strongest dependencies between the fuzzy output and the inputs. This approach takes advantage of predictive capabilities of FNN and gives reasons to explain the output and the patterns discovered by the IKM part of the proposed system. It also highlights and adjusts some explicit rules given by human experts.

A pre-processing step to delete the contradictory rules was done. We considered as contradictory the rules: (1) those having different output predictions than the same input class, and a relative small coefficient of trust:

```
IF RdaFit1 is:Medium THEN class is:VeryLow (47.79%)
IF RdaFit1 is:Medium THEN class is:High (47.62%)
```

or (2) rules showing big differences between the input (the classification) and the output of the system (the inputs and the outputs must be in similar domains):

```
IF KnnXFi1 is:VeryLow THEN class is:Low (54.98%)
IF KnnXFi1 is:Low THEN class is:VeryLow (55.99%)
IF KnnXFi1 is:High THEN class is:Low (78.70%)
```

Finally, the following list of the trustiest fuzzy rules was considered for the chosen net (IKM-FNN10H):

```
IF CarFit1 is:VeryLow THEN class is:High (39.22%)
IF CarFit1 is:Low THEN class is:High (82.30%)
IF CarFit1 is:Medium THEN class is:High (48.74%)
IF CarFit1 is:High THEN class is:High (39.04%)

IF SimFit1 is:VeryLow THEN class is:Medium (61.25%)
IF SimFit1 is:Low THEN class is:Medium (36.04%)
IF SimFit1 is:High THEN class is:Medium (43.72%)

IF RdaFit1 is:VeryLow THEN class is:Low (75.65%)
```

```
IF RdaFit1 is:Low THEN class is:Low (100.00%)
IF RdaFit1 is:High THEN class is:High (76.39%)
```

Three types of fuzzy rules were obtained (see above): (1) some could be grouped by the same output, or (2) by having the same fuzzy term in the premise and conclusion, and, finally, (3) rules with mixed terms in premises and conclusion parts. From the first two groups of fuzzy rules (in italics), we could conclude that the opinion of the specific entry classifier is not important for the given output. More precisely, it is better not to consider the CART prediction for values *High* of toxicity (class 4) when comparing with the other approaches.

```
IF (CarFit1 is:VeryLow) OR (CarFit1 is:Low) OR (CarFit1
is:Medium) OR (CarFit1 is:High) THEN class is:High
```

Similarly, SIMCA is not significantly important for predicting class 3 (*Medium* values of toxicity: the second group of fuzzy rules). From the last group of rules (in bold), we could find the best classifier: to predict class 2 (*Low* toxicity) is better to consider the opinion coming from RDA. The same opinion is very important for predicting the class 4 (*High* toxicity) cases too.

Our approach showed an improved behavior as a combination of classifiers. The model is viewed as a specific part of an hybrid intelligent system, in which different knowledge representation approaches could improve the final model. The results on fuzzy rules extraction, and the possibility to interpret particular inferences, suggest the neuro-fuzzy approach has the potential to significantly improve classification methods used for toxicity characterization.

6.5.4 Conclusions

Classification of the toxicity correlated to the descriptors for organic compounds requires a high degree of experience from computational chemistry experts. Several approaches were described to generate suitable computer-based classifiers for these patterns. The experts range from a QSAR equivalent FIS, to classical ANN architectures, through to neuro-fuzzy nets. The main problem regarding the symbolic approach is the difficulty of improvement, due to the existence of limitations in knowledge elicitation. Several implicit knowledge models with different hidden layers were designed, trained and analyzed using.

Advantages of developing HIS-based models to combine implicit and explicit knowledge as neural and neuro-fuzzy rule-based structures were identified. Considerations about how to improve the structure and the behavior of proposed models were also discussed, in order to extract compact and comprehensible set of rules about the problems described. The proposed models exhibit effective solutions. For example, the model of combination of classifiers follows an opposite philosophy of traditional selection approach, in which one evaluates the available systems against a representative sample and chooses the best combination of available methods.

Our study intends to contribute to the understanding of the possibilities to represent the knowledge about the hazardous waste toxicity (Mazzatorta et al. 2002, 2003). The presented approaches are case studies based on hybrid intelligent systems, combining artificial neural networks (ANN) and QSARs, on the basis of neuro-fuzzy modules implementation. The proposed neuro-fuzzy knowledge representation gives an encouraging alternative to the stochastic models. We proved such models are able to learn sophisticated collections of descriptors of industrial organic compounds. In addition, they are able to represent knowledge acquired from human experts in order to improve the prediction results. The important features of the proposed models are certain abilities to generalize and adapt to noisy data, and algorithms to improve overall prediction by combining various existing experts' opinions.

Acknowledgements

The work presented in Sect. 6.5 was partially funded by the EU FP5 Research Training Network IMAGETOX (Intelligent Modelling Algorithms for the General Evaluation of TOXicities http://airlab.elet.polimi.it/imagetox) under the contract HPRN-CT-1999-00015 and EU FP5 Quality of Life project DEMETRA (Development of Environmental Modules for Evaluation of Toxicity of pesticide Residues in Agriculture http://www.demetra-tox.net) under the contract QLK5-CT-2002-00691. The case studies are based on collaborative work and data made available by Dr. Emilio Benfenati, Dr. Mark Cronin, Dr. Aynur Aptula, Dr. Tatjana Netzeva, Prof. Terry Schultz and Prof.ssa Giuseppina Gini. The author, Dr. D. Neagu, thanks all of them for continuous support and expertise.

7 New Trends of Developing Hybrid Intelligent Systems – AIS Hybridization and DNA-Hybridization

As already mentioned in Chap. 1, the use of biologically inspired *CI* techniques play a crucial role for the hybridisation at any level of *HIS* features and performances. Two emerging and promising biologically inspired techniques seem to be the impulse of the moment in the field of *HIS*, namely: **A**rtificial **I**mmune **S**ystems (*AIS*) and *DNA* computing. The computing framework featuring each of theses two above-mentioned techniques was enlarged by original and efficient hybridisation methods. A lot of interesting real-world applications are reliant on *AIS* and/or *DNA* techniques despite the fact that these methods are still at their incipient stage. Their applications cover such different areas as bioinformatics, biological engineering and biologically motivated computing. The interdisciplinary development approach of *AIS* and *DNA* reliant hybridisation algorithms, techniques and application was created by the collective effort of a large spectrum of high technology practitioners; mainly the computer scientists, engineers acting in different technical fields, biologists and natural environment specialists.

7.1 A Brief Introduction to AIS Computing Framework

Because of its strong intrinsic hybrid structural features, *AIS* is, perhaps, the most suitable *CI* technique for hybridisation. A general definition for an *AIS*, as proposed in (de Castro and Timmis 2002) but emphasising a strong connection to *HIS* is stated as follows: *AIS* are adaptive (computational) systems inspired by theoretical immunology – from the vertebrate immune system, especially – and by observed immune functions and principles which are applied to typical problems that *HIS* can solve.

Both a suitable modelling and an effective exploring of *AIS* requires some fundamental knowledge regarding the structure and functions of vertebrate immune systems – at different levels (cells, molecules, organs). It is beyond the aim of this book to deliver this immunological background; see for this (de Castro and Timmis 2003), where this task is well done from the engineering perspective of an application oriented information processing. Also for AIS understanding in context of *HIS*, namely for AIS hybridisation with other *CI* techniques (*NN, FS, EC, KBES*), some elements of knowledge are required

Mircea Gh. Negoita, Daniel Neagu, and Vasile Palade: *Computational Intelligence: Engineering of Hybrid Systems*, StudFuzz **174**, 151–161 (2005)

regarding the Immune System connections with other biological systems such as the nervous, endocrine, evolutionary biological and the cognitive systems.

The main structural component elements of *AISs* used in immune engineering are the representation of operative elements, the operators and the algorithm modelling the dynamics of system behaviour.

The operative element (the individual in a population) is called an *antibody* (it recognizes and binds to a specific molecular pattern) and the *antigen* is the corresponding molecular pattern recognized by a specific antibody. An antibody binds to many antigens, but its degree of binding is variable with respect to which antigens. The evaluation of this recognizing power is qualitatively featured by the *affinity* measure (usually, a distance measure) between the (string) representation of the two operative elements, an antibody and any of the recognized antigens by this antibody. The concept of affinity is equivalent to the fitness function of the chromosomes in a *GA*. The space of individuals (string representations) is called the *shape-space* and the region located in the shape space where the population of recognized antigens by a particular antibody is sited, is called a *volume* (de Castro and Timmis 2003). A point in the space of individuals may be represented by a string of real values, integers, bits or symbols. The selection of the most suitable representation is strongly dependent on the application being solved.

Some of the immune algorithms modelling the dynamics of *AIS* system behaviour are as follows: negative selection, positive selection, clonal selection and immune network models.

The *Negative selection* algorithm acts on a population of antigen individuals by monitoring changes on them so that the antibody individuals of another population (the detectors) can be continually matched against these antigens. This algorithm is typically reliant on individual bit matching of antibodies against antigens, both kind of individuals being binary represented (Gonzales et al. 2003). Different versions of the negative selection algorithm are using different rules of individual bit matching of binary represented detectors and antigens. If an antibody (detector) string fails to match an antigen string, it is retained for another detection (iteration); otherwise it is "negatively" selected (this means rejected or eliminated).

The *Positive selection* algorithm selects the antibodies that do not recognize any antigen and place them in a set of unmatched antibody individuals. For each of these unmatched antibodies an *AIS* operator – guided mutation for example – is applied for increasing the degree of antibody matching (affinity) with the unrecognised individuals of antigen population that it best matches. When its degree of matching is above a threshold, that antibody is (positively) selected in a set. If its affinity is still below the threshold, that antibody is kept in the population of unmatched elements and the *AIS* operator is again iteratively applied to each of these antibodies. The stop condition of positive selection algorithm is when a pre-determined number is reached of matched antibodies (de Castro and Timmis 2003).

The *Clonal Selection* algorithm is inspired by coping process of the immune system as it deals with invader organisms. There are different opinions, resulting in different models of clonal selection (Forrest et al. 1993; de Castro and von Zuben 2000; Castro and Timmis 2003). But the essential computational framework of clonal selection involves the same steps. By this kind of selection, an antigen individual selects a number of immune antibody cells for cloning (proliferation), the proliferation being proportional degree of matching (affinity to) the external selective antigen. The dynamics of the clonal selection algorithm evolves the *AIS* from an initially static state, in absence of any external stimulus ("invading" antigen) to a set of active states responding to the presence of external "invaders". Another aspect of *AIS* dynamics is the reproduction, under clonal selection of antibodies where the *AIS* (mutation) operator acts less intensively on antibodies with the highest affinity of matching the selective antigen and more intensively on those with a lower affinity. The diversity of the antibody population is kept by inserting a particular number of random individuals at the end of each algorithmic iteration (Kelsey and Timmis 2003).

Different *Immune Network Models* are conferring a common feature to AIS: behaviour reliant on a permanent dynamic activity, even in the absence of any external stimulus ("invading" antigen). These immune network models differ as to their antigenic representation by hoe they use their AIS operators or algorithms. These algorithms are responsible for clonal selection, expansion (reproduction) and interactions of the high affinity antibodies, affinity reliant maturation (the mutation operator acting on the antibody clones), construction, interaction and/or suppression of network antibodies. Two original Immune Network algorithms are described in detail and their implementation by pseudocode is presented in (de Castro and Timmis 2003): *RAIN* (*Resource limited* **A***rtificial* **I***mmune* **N***etwork*) and *aiNet* (**A***rtificial* **I***mmune* **Net***work*). Real-valued vectors were used in the original implementation of these two Immune Network algorithms. *RAIN* produces a topological representation of antigenic patterns, a favourable framework for the identification of clusters and inter-relationships between data items. *aiNet* is relied on the reproduction operator of a discrete set of antibodies where the reproduction is guided to follow the spatial configuration of the antigenic landscape.

7.2 A Survey on AIS Application to Hybrid Intelligent Systems

Artificial Immune Systems are a new *CI* approach that has both its own standalone applications and applications involved in the *HIS* framework. Because of the intrinsic hybrid systems nature of *AIS*, practitioners are not surprised that applications are being implemented within the *HIS* framework.

An introduction is made in (Ishida 2004) to the *AIS* as an information system, as a network, as an adaptive system and as a self-defining system. This book is an engineering framework delivered to practitioners for implementing the *AISs* in different applications, but more fôcused on the field of diagnosis and control.

The *AISs* were proved to be superior to hybrid *GA* in function optimization (Kelsey and Timmis 2003). Here the *AIS* algorithm inspired by clonal selection and called *BCA* (**B**-cell algorithm) got a high quality optimisation solution by performing significantly fewer evaluations than a *GA*. A unique mutation operator was used – contiguous somatic hyper mutation – that operates by subjecting the contiguous region of the operative element (vector) to mutation. Instead of multiple random sites being selected within the operative *AIS* element, only one random site is selected within this vector, with a random length. The vector is mutated until the randomly selected length of the contiguous region is reached. See Fig. 7.1. The random length utilised by this mutation operator confer to the *BCA* individual the ability to explore a much wider region of the affinity (fitness) landscape than just the immediate neighbourhood of an individual. A hybrid clonal selection *AIS* was used more successfully than the evolutinary algorithms for solving a combinatorial optimisation application, the Graph colouring problem (Cutello et al. 2003). Here the use of a crossover operator was avoided by using a particular mutation operator combined with a local search strategy. In this way there was no embedding specific domain knowledge.

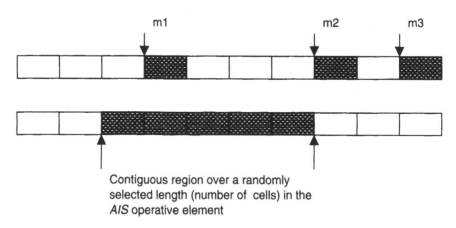

Contiguous region over a randomly
selected length (number of cells) in the
AIS operative element

Fig. 7.1. The difference between usual-punctual-mutation (*top*) and contiguous mutation operator (*bottom*), after (Kelsey and Timmis 2003)

AISs that are based on clonal selection have proven to be effective both for combinatorial optimisation and for machine learning problems. An *AIS* modelled as a noisy channel has applicability in adaptive noise neutralization (Cutello and Nicosia 2003): the signal is the population of B-cells, the channel is the global *AIS*, the noise source is the antigen and the received signal E is the antibody.

Pattern recognition is another application area of *AIS*. An Idiotypic Network Model using immune system antibodies as partial templates performed the feature extraction in pattern recognition of Chinese characters that have a common sub-structure in the character. This model contains the effect of diffusion of antibodies. The amount of diffused antibodies is calculated by adopting the (spatial) distribution of antibodies centroids as the virtual points where antibodies were concentrated and redistribution of antibodies is performed accordingly (Shimooka and Shimizu 2003). A very interesting **M**ultilevel **I**mmune **A**lgorithm (*MILA*) was proposed for novel pattern recognition and was tested with anomalous pattern problems (Dasgupta et al. 2003). The algorithm has four phases: Initialization phase, Recognition phase, Evolutionary phase and Response phase. Real-valued strings represent both antibodies and antigens, which is different to the Negative and Clonal selection. The antigen/antibody recognition is modelled by using a Euclidian distance measure as the degree of matching. The anomalous patterns are detected by two measures of effectiveness:

$$\text{Detection rate} = \frac{TP}{TP + FN} \; ; \quad \text{False alarm rate} = \frac{FP}{TN + FP}$$

where,

TP (**T**rue **P**ositive) – stands for the anomalous elements correctly identified as anomalous; TN (*True* **N**egative) – stands for normal elements correctly identified as normal; FP (**F**alse **P**ositive) – stands for normal elements erroneously identified as anomalous; FN (**F**alse **N**egative) – stands for anomalous elements erroneously identified as the normal. The parameters of the *MILA* algorithm are controlling different detector thresholds that change, so that different detection rates and false alarm rates are obtained.

Security systems for computers and the Internet work environment are another productive application area of *AIS*. Intrusion detection on the Internet – for internal masqueraders mainly, but for external ones too – is reported in (Okamoto et al. 2003). The method was inspired by the diversity and specificity of an immune system, namely each immune cell has a unique receptor featured by a high degree of matching a specific antigen. So each of the agents that are used in this approach has its unique profile and computes a high score against the sequential (command) set typed by the specific user matching this profile. By evaluating all the scores (for all the profiles), one of the agents determines whether a particular user is an intruder (masquerader) or not. *AIS* self-monitoring approach has applicability also in solving specific problems to distributed intrusion detection systems (Watanabe and Yshida 2003). Usually computer protection is referring to anti-virus protection and to intruders' detection, but (Oda and White 2003) applied *AIS* for immunity to unsolicited e-mail (spam or junk mail) where regular expressions, patterns that match a variety of strings, are used as antibodies. These regular expressions are grouped in a library of gene sequences and in their turn the regular expressions are combined to randomly produce other regular expressions to produce antibodies that match more general patterns.

Many other *AIS* applications have been developed in other areas such as: autonomous navigation, computer network security, job-shop scheduling, data analysis and optimisation (de Castro and Timmis 2003). Some of these applications are reliant on the idea of combining *AIS* with different other *CI* techniques (*FS*, *NN*, *EA*, *KBES*, *DNA* computing) with the aim of creating hybrid intelligent systems that are collecting the individual strength of each *CI* component.

An *Immune Multi-agent NN* was introduced in (Oeda et al. 2003) that has a high classification strength reliant on a multi-agent framework where each immune agent relies of its own *NN* learning and distributes to exchange their *NN* stored information. The *NNs* used are of a sand glass type: the input and output layer have the same number of neurons and the *NNs* are trained by back-propagation to calculate output activities that are the same as input patterns. The hidden layer of these *NNs* has the property of storing the condensed information after the training is finished. The information in the hidden layer is "condensed" because the number of neurons in this layer is less than that in the input/output layers. This *HIS* in the form of an immune multiagent NN was applied in the medical diagnosis of hepatobiliary disorders by classifying a database of this disease with negative examples. Another *HIS* relying on *NN-AIS* hybridization consists of a competitive learning *NN* that is featured by simultaneously iteratively weight updating and the automatic generation of a new optimised *NN* architecture (de Castro and Timmis 2003). The *NN* weights are antibodies that iteratively match more perfectly against antigens under a hypermutation operator that is equivalent to a multi-point *GA* mutation.

Another *HIS* example is reliant on a *EC-AIS* hybridization – a framework that handles the *GA* constraints by using immune techniques for generating structural *GA* elements (niche, species, populations). For example, a job-shop scheduling application (Hart and Ross 1999) was solved by a two phase *HIS* that fast generates new schedules as a result of external perturbations in the production environment as follows: a *GA* reliant phase (phase 1) – a *GA* acts to select the common patterns (gene segments) among highly frequent scheduling sequences used in the production process; an *AIS* reliant phase (phase 2) – the new schedules are generated by an *AIS* acting on the gene libraries created in phase 1.

FS-AIS hybridization was strongly used in a wide range of real-world applications. The interdisciplinary field of robotics and advanced autonomous control problems. For example autonomous aircraft control application are a typical application field of these kinds of *AIS*. Distributed Autonomous Robotic Systems may have a *FS* like modeling of the stimulation level of an antibody (individual robot strategy of action) while an *AIS* relying on clonal selection is used for transmitting high quality strategies of action (antibodies) among the robots. No central control exists and the role of antigens is played by elements of external environments (Jun et al. 1999).

The *KBES* intelligent hybridization with *AIS* is known especially by immune inspired case-based information organization and retrieval: the case-based *AIS* is in form of a case memory that handles the antigens, new cases (problems) to be solved, and the antibodies – previous (past) cases in the system (Hunt et al. 1995).

DNA-AIS hybridization will be briefly introduced in Sects. 7.3 and 7.4.

A recent intelligent hybridization of *AISs* is applied in case of one of the most revolutionary technology nowadays, namely in case of **E**volvable **H**ardware (*EHW*). A main reason for *EHW-AIS* hybridization was reliant on two *AIS* features, healing and learning, that were applied to design *EHW* fault – tolerant *FPGA* systems (Bradley et al. 2000). An additional layer that imitates the action of antibody cells was incorporated to the previously elaborated embryonic architecture by the same team (Ortega et al. 2000). Two variants of this new *EHW* architecture use an interactive network of antibody cells featured by 3 independent types of communication channels: the data channels of the embryonic array of cells, the data channels of antibody array of cells and the inter-layer communication channels ensuring that antibody cells can monitor the embryonic cells. The antibody array of cells performs monitoring and checking of the embryonic array of cells, so that the correct functionality of any particular *EHW* configuration can be achieved at any time. Another *AIS* inspired variant of *EHW* hardware fault detection was reported in (Bradley and Tyrrell 2001). They used an advanced *FPGA* hardware-Virtex XCV300 – to implement a hardware negative clonal selection *AIS* attached to a **F**inite **S**tate **M**achine (*FSM*). This is very important because any hardware system can be represented by either a stand-alone or an interconnected array of FSMs.

7.3 A Brief Introduction to Fundamentals of DNA Computing

One aspect of interdisciplinary interaction of biology, electronics engineering and computer science led towards research aimed to develop techniques involving information processing capabilities of biological system to supplement, and, perhaps, finally to replace the current silicon-based computers by so called *Quantum Computing* and *DNA Computing*.

Some basic *DNA Computing* vocabulary, terms and abbreviations, as from (Paun et al. 1998):

1. *DNA - DeoxyriboNucleic Acid*
2. *A DNA molecule* – is a structure of polymer chains usually referred to as *DNA* strands. *DNA strand* is the data structure employed in *DNA* computing
3. *Nucleotides* – are the structural elements of a strand. The nucleotides may differ only by their *bases.*
4. There are four kinds of *bases: A (**a**denine), G (**g**uanine), C (**c**ytosine), T (**t**hymine).*
5. *A double helix of DNA* – results by bonding of two separate strands. Bonding is the result of a pair wise reliant attraction of bases under the following rules: an *A* base always bonds with a *T* base; a *G* base always bonds with a *C* base

Two basic features of the strands are crucial for the computational power of *DNA* computing (Paun et al. 1998):

1. *DNA* strands are structured by a large number of basis (high density of information). This *high structural parallelism* made them suitable for powerful parallel computing
2. The bonding principle – known as *Watson-Crick complementarity* – facilitates different effective information encodings and also different operators on *DNA* strands, that give another strength to DNA computing

The aim of this book is just to give a general view on *DNA* computing hybridisation with other Intelligent Technologies. The complete mathematical foundation of *DNA* computing and a detailed description of typical *DNA* computing operators acting on *DNA* strands are to be found in (Paun et al. 1998). We will just limit ourselves to mention very briefly some of the main operators that are typical to *DNA* computing as from the above named reference book.

Denaturation is the operation separating a double stranded *DNA* into its two strands without altering (breaking) the single strands.

Renaturation is the operation fusing two separated strands. This operation is the reverse of denaturation and is performed under principle of *Watson-Crick complementarity.*

Lenghtening is the operation of adding some basis – a chain of basis called the *template* – to an existing *DNA strand.*

Shortening is the operation of removing basis one at a time from the ends of a *DNA* strand.

Splicing – is the operation of cutting two strands, followed by a concatenation that is performed cross wisely; the cutting points are specified by given sub strings.

Extraction – is an operation performed on a set (tube) of strings (strands) having a reference sub string as the separation reference. All strands which contain the reference as a (consecutive) string in the tube are separated from the all strands in the tube which do not contain the reference as a sub string.

Cutting is the operation of destroying internal bonds in a DNA strand. There are different variants of this operation as to what, where and how the cutting is made.

Pasting is the operation of linking different DNA strands. This operation can be performed in different ways. *Amplification* – is the iteratively operation of exponentially multiplying a *DNA* strand (the target) when the flanking sequences (called borders) of the target are known: after n iterations, 2^n samples of the initial *DNA* strand are got.

Detect and *Length-separate* is the operation of detecting the presence of *DNA* strands, together with a separation of all *DNA* strands featured by a length less or equal to a reference length.

DNA computing called also **Molecular Computing** (*MC*) utilizes a framework of rules inspired by interaction of biomolecules and the protocols that govern microbiology. But developments are still far from being reliable and easy to implement in silicon-based computers. A software platform *Edna* – has been developed to improve the encoding, reliability and efficiency problems of molecular computing (Rose et al. 1999). An important and useful component of this software platform is the *virtual test tube* (a test tube simulator) giving practitioners a practical tool for a realistic test of any *DNA* computing protocol, before its implementation in a real Lab test tube.

7.4 A Survey on DNA Applications to Hybrid Intelligent Systems

DNA computing hybridisation with *EC* was applied to very practical purposes: *DNA*-like genomes proved to be advantageous for *GA in silico* as reported in simulation on the Hamiltonian Path Problem (West et al. 2003). The solution of this problem is directly applied in the field of combinatorial optimisation, as for route planning or network efficiency. *DNA*-like genomes representation in this *GA* means that each base in the *DNA* strand is a single character and the entire strand is a string, which may contain single or double stranded sections, bulges and other secondary structures. A population of partially formed potential solutions is evolved. The individuals in this population could react in a pre-established manner with other individuals (partial solutions) to iteratively form fitter solutions. The reaction operator applied during the successive iterations is called *extension* since longer strands are formed ultimately representing all possible paths in the digraph representing the instance of the problem. Several on line fitness functions (*promise* fitness, extension fitness, *demand* fitness, *repetition* fitness) are used with the aim that the new individual (potential solution) inherits the good features of the partners in reaction (under the extension operator). This hybrid *GA* with *DNA* has two kinds of advantages over the classical *GA*. These computational advantages are highly parallel and distributable operation, noise immunity due to its asynchronous computational mode, and, cost and time savings for small clusters of conventional sequential computers.

Another example of *DNA-GA* hybrid intelligent system was used for a sequence alignment method (Vallejo and Ramos 2003). A finite state automaton generates a collection of two-dimensional maps of *DNA* sequences. These two-dimensional maps are overlapped one over another with the aim of discovering coincidence in character patterns. The *GA* is used for getting the optimal overlapping by evolving the overlapping Cartesian positions of a map over a reference plane.

Molecular *GP* is another sample of DNA computing hybridisation with *EC* (Wasiewicz and Mulawka 2001). This new *GP* method is suitable for dataflow computers and is applied on the populations of graph-structures describing logical functions. The input data are arguments of logical functions, therefore input signals to the data flow function graph. The function graph structure is made of special and arc *DNA* strings. *DNA* computing operators as renaturation, denaturation, cutting, concatenation and amplification are executed during different steps of the GP algorithm, as the case dictates. Amplification is performed just only in one test tube.

Aspects of DNA-*FS* intelligent hybridisation are introduced in (Deaton and Garzon 2001). As is to be expected such systems hybridisation relies on the fact that the intrinsic uncertain and inexact nature of the chemical reactions governing *DNA* computation can be exploited with the aim of implementing robust soft computing systems. A system of *DNA-FS* intelligent hybridisation tries to overcome the over idealization aspects of *DNA* computing. A fuzzy variable can be implemented as a species of *DNA* strands. Starting from the idea that, in general, a given *DNA* strand can be represented as a linear combination of some independent *DNA* sequences, a desired fuzzy membership function is built by designing a set of encoding *DNA* strands as follows: one *DNA* strand will represent the complement of the input *DNA* strand that exhibits the maximum membership value; another *DNA* strands represent other membership values by lesser or greater degree of renaturation. All of these DNA strands approximate the values of the point wise membership function. The same work proposed an implementation method of fuzzy rules association (fuzzy inference) in *DNA* renaturation on a *DNA* chip, relying on the technology called *DNA* computing on surfaces. Fuzzy logic with biomolecules confer the advantage of *DNA* computing that is more tolerant to errors and is more honest to the natural reality.

An *HIS* that is *based* on combining *DNA* with *NN* is discussed in (Mills et al. 2001). A DNA *Analog NN*, namely a Hopfield *NN*, was implemented by representing the elements of memory in this *NN* in form of a set of single stranded *DNA* sequences under *DNA* operations as renaturation, pasting and cutting via a restriction enzyme.

A *DNA-AIS* hybrid intelligent system was reported in (Deaton et al. 1997). An *AIS* negative selection algorithm was implemented in a *DNA* computing framework. Namely, the censoring and monitoring parts were by using *DNA* single strands under denaturation, renaturation and splicing operators.

8 Genetic Algorithms Based Hybrid Intelligent Systems

The oldest branch of Evolutionary Computation, namely *GA*, is at the same time the used in real-world applications of *HIS*. Chapter 8 of the book tries to introduce some of these applications. The introduced applications were selected so as to illustrate the efficiency of *GA* used in combination with other intelligent technologies.

8.1 GA Based Hybrid Intelligent Systems in Fuzzy Information Processing

A typical example of intelligent techniques interacting in the *HIS* framework is *GA* hybridization with a *FS*.

GAs are applied in *FSs* for tasks such as learning the fuzzy *IF-THEN* rules, design of fuzzy controllers, or the elaboration of some learning methods that are based on fuzzy reasoning by using *GA* in combination with descent methods.

8.1.1 GA as Tools to Design Fuzzy Systems

GA hybridization with *FSs* may be used for the *FS* application in *GA* design (see Fig. 8.1), with different purposes, as the case is, for so-called *Fuzzy Genetic Algorithms (FGA)*.

FGA There are several types of *FGA*. For example, *FGA* with fuzzy fitness functions in two variants:

- The fitness function is directly delivered (by the operator) and naturally, subjectively estimated (by fuzzy membership grades)
- The fitness function is indirectly delivered (by the fuzziness of predicates related to chromosomes)

This type of *GA* has typical applications like the inductive learning of *IF-THEN* rules for fuzzy controllers or fuzzy knowledge bases.

Another example is when *FGA* is used because of codification *with* two variants also:

Mircea Gh. Negoita, Daniel Neagu, and Vasile Palade: *Computational Intelligence: Engineering of Hybrid Systems*, StudFuzz **174**, 163–194 (2005)
www.springerlink.com
© Springer-Verlag Berlin Heidelberg 2005

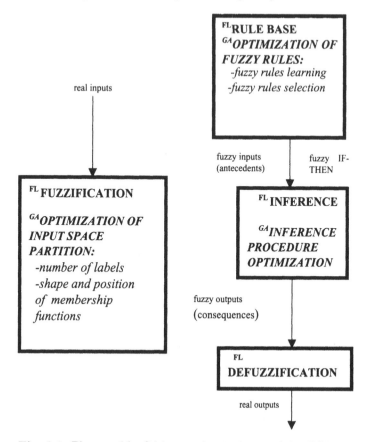

Fig. 8.1. The possible *GA* interaction in design of the *FS* block

- with chromosomes being the result of codification of a fuzzy set (or a lot of suitable fuzzy sets for concatenation)
- with chromosome obtained by codification of a fuzzy relation, represented by a matrix in the finite case

A typical application for this last type of *FGA* is the optimization role in systems modeling with a fuzzy relational equation represented by sup-min compositions (Negoita 1994; Negoita et al. 1994).

8.1.2 GA Based Hybridization for Knowledge Handling in Expert Systems (Fuzzy Rules Learning and Selection)

GAs solve two basic problems of knowledge bases (Negoita et al. 1994b):

- knowledge base building (fuzzy rules learning)
- knowledge filtering (fuzzy rules selection)

Knowledge Learning by GA. The knowledge base is built in the following manner: given a set of crisp examples, inputs-output, that are experimentally obtained, the GA searches for a better combination of fuzzy antecedents and consequences of the rules according to the set of examples. The fuzzy input variables X_i are triangular with X_i^m fuzzy labels; the Y output variable has Y_j singleton labels. In experiments the example $i = 2, j = 3, m = 3$ was used. The rule has the form:

IF X_1 is $\left\{ X_1^1 or X_1^2 or \dots X_1^P \right\}$ **AND** $\dots$ **AND** X_i is $\left\{ X_i^1 \text{ or } X_i^2 \text{ or } \dots X_i^P \right\}$

THEN Y is Y_j.

The potential solution of the problem, the set of fuzzy *IF-THEN* rules, is encoded as a chromosome consisting of a matrix $A_i(3 \times 9), i = 1, 2, 3, 4$. The method of encoding the elements in a chromosome, in our case the possible antecedents of the rule. As follows: 1 if the fuzzy label is a present value to the respective variable antecedents of the rule; 0 otherwise. Individual matrices compose the population A:

$$A = [A_1 A_2 A_3 A_4 A_5 A_6]^T , \quad A(18 \times 9) .$$

The matrix of examples is $E(1 \times 3)$.
The fitness function for a chromosome is defined as follows:

$$Ev(A_i) = \sum_{k=1}^{l} |z(k) - e(k, 3)|$$

where $z(k)$ is the result of fuzzy inference engine (*max, min, truncation*) of fuzzy *IF-THEN* rules on (A_i) on the example k, and $e\ (k, 3)$ is the consequence in line k of E. The evaluation on an example k, $z(k)$, is of the form:

$$z(k) = \sum_{j=1}^{m} \lambda_j . c_j \sum_{j=1}^{m} \lambda_j ,$$

where c_j is the consequence in the rule j of A_i and $\lambda_j = \min(\max\{\mu_L{}^1, \mu_M{}^1, \mu_B{}^1\}, \max\{\mu_L{}^2, \mu_M{}^2, \mu_B{}^2\}$, where $\mu_L{}^1, \mu_M{}^1, \mu_B{}^1$ and $\mu_L{}^2, \mu_M{}^2, \mu_B{}^2$

are the membership grades corresponding to the example k on the universe of discourse of the input variables X_j in the rules of A_i. The membership degrees are weighted by the corresponding antecedents values in the rule.
The GA starts with the evaluation of the initial population followed by setting the elements of population in a decreasing order. The vector $E_{val}(i')$, $i = 1, 2, 3, 4$ is created with:
$A_{1'} = E_{val}(1)$, $A_{2'} = E_{val}(2)$, $A_{3'} = E_{val}(3)$, $A_{4'} = E_{val}(4)$, and the population is reordered:

$$A = [A_1, A_2, A_3, A_4, A_5, A_6,]^T .$$

If $E_{val}(1) \leq \boldsymbol{THRESHOLD}$, then the GA is stopped and A_1 is the optimal solution. $\boldsymbol{THRESHOLD}$ is the optimum value for the fitness function.

If $E_{val}(1) > \boldsymbol{THRESHOLD}$, then the GA starts by chaining the crossover-type and mutation-type operations. An original $MATLAB$ created function calculates a probabilistic vector and probabilistically allows as the case:

- the choice of operation (crossover and mutation)
- the choice of the first parent for both the mutation and crossover
- the choice of the second parent in case of crossover
- the configuration of template for crossover

Each iteration is finished with an evaluation followed by an ordering, as with the initial population.

A special matrices operating crossover operation is defined to produce offspring of the following form:

$A_5 = min(A_{r1}, T') + min(A_{r2}, T)$
$A_6 = min(A_{r1}, T) + min(A_{r2}, T')$ where
A_{r1}, A_{r2} are the parents matrixes
T' is the complentary of T
$T(3 \times 9)$ is the template matrix introduced for crossover

The $STOP$ conditions are as follows:

- at any iterations, in case of $E_{val}(A_1) \leq \boldsymbol{THRESHOLD}$ (that means A_1 is the optimal solution)
- at any H_{max} iterations in one of the V_{max} cycles if GA doesn't progress (the evaluation function doesn't decrease at any inside iteration during consecutive two V cycles) – that means GA stops and indicates the solution stagnates and prints the stagnated solution.

In case of starting with an initial population A as follows:

A

```
0 1 1 0 0 1 1 0 0          the optimal solution A_i is
0 1 1 1 1 0 0 1 0
1 1 1 1 1 1 0 0 1            1 0 1 1 0 0 1 0 0
1 0 1 1 1 0 1 0 0            1 1 1 0 0 1 0 1 0
1 1 0 0 0 0 0 1 0            0 1 0 0 1 0 0 0 1
0 0 1 1 0 1 0 0 1
1 1 1 0 1 0 1 0 0            after V = 4 cycles at
1 1 0 0 1 0 0 1 0            at generation h = 115
0 1 0 0 1 0 0 0 1              in the 4th cycle.
0 0 1 1 1 0 1 0 0
1 1 0 1 1 1 0 0 1
0 1 0 0 0 1 0 0 0          The initial number of cycles
0 1 0 1 0 0 0 0 0             V_max = 10 of h = 200
1 0 1 0 1 1 0 0 0
1 0 1 1 1 1 0 0 0
```

More details and results on the convergence of this *GA* are to be found in (Negoita et al. 1994b).

Knowledge Filtering by GA. The GA finds the best set of *IF-THEN* rules. The chromosome C is a string of n bits: 1 if the rule is esteemed to be eligible, 0 otherwise and n is the total number of rules: $C = \{0, 1, 0, \ldots, 0, 1, 1, 1\}$. The example matrix is the same as in the previous paragraph.

The evaluation function is defined as follows:

$$E_v(C) \sum_{p=1}^{l} = E(p) = \sum_{p=1}^{l} |e(p, 3) - \sum_{j=1}^{m} \lambda_j c_j \Big/ \sum_{j=1}^{m} \lambda_j| = \sum_{p=1}^{l} |e(p, 3)$$
$$-z(p)|, \text{ where}$$

$m \leq n, m$ – the number of eligible rules
$z(p)$ – the result of inference due to the rules with weight 1 in C.

If the matrix of rules (the knowledge base) contains the whole possible combination of *IF-THEN* rules, this *GA* becomes the generalization of the one presented in the previous paragraph. The set of best rules may contain more rules in this case with the same consequence and the *GA* becomes more efficient, at the same time using more simple chromosomes (single strings of bits, like the canonical *GA*), the classical *GA* operators being in the simplest form. During the *GA* evolution, the partial overlapping of successive generations was used. The only limitation of this *GA* is in the case of a large knowledge base. This is because there is an exponential increasing in the dimension of the searching space as a function of the number of fuzzy labels of both antecedents and consequences.

The *GA* for knowledge learning was tested in a manner suggested in (Herrera et al. 1993), namely: given a set of rules, a set of uniform distributed examples in the field of input data and running the *GA* on this set of examples with a randomly chosen initial population. The set of rules used in generating the examples was finally obtained again. The *GA* led to the initial given rules.

8.1.3 GA Based HIS for Solving Relational Equations

Many *FS* applications intensively use *fuzzy relational equations*, namely the same combination of matrix entries as in the regular matrix multiplication, but *fuzzy set intersection* and *union* are the used operations instead of *multiplication* and *addition*. *GA* is a useful tool for finding approximate solutions to a system of fuzzy relational equations. As mentioned in (Cordon et al. 1997), a reference work in this kind of *HIS* applications is (Negoita et al. 1994c). The problem was concerned with solving a system of fuzzy relational equations of the following type:

$$X_1{}^\circ R = Y_1$$

$$\cdot$$
$$\cdot$$
$$\cdot$$

$$X_K,{}^\circ R = Y_K, \text{where:}$$

"$\circ$" stands for $\max - \min$ composition ,

X_k, Y_k are the fuzzy sets defined on finite universes of discourse, namely:

$$X_k \in F(X), \quad X = \{x_1, \ldots, x_n\} ,$$
$$Y_k \in F(X), \quad Y = \{y_1, \ldots, y_m\} .$$

$F(.)$ denotes the family of fuzzy sets, while R stands for a fuzzy relation defined in the Cartesian product $X \times Y$, $R \in F(X \times Y)$. The system of equations has to be solved with respect to R, treating all pairs (X_k, Y_k) as provided.

Under the assumption that the system has a non empty set of solutions, the fuzzy relational equation theory proves (Sanchez 1970), that *a greatest solution*, say R, is taken as the intersection of α-composition between respective fuzzy sets (X_k, Y_k), i.e.:

$R = \cap_{k=1, K} R_k$, where

$R_k = X_k \alpha, Y_k$ is the greatest solution of the "k" equation $(X_k{}^\circ R = Y_k)$ and "α" stands for the "α"-composition, namely:

$$\text{for every } a, b \in [0,1] \qquad a\alpha b = 1, \quad \text{if } a \leq b$$
$$b \quad \text{elsewhere.}$$

The above stated assumption is poorly satisfied in most cases, so the following situation was considered: each individual equation "$X_k{}^\circ R = Y_k$" of the system has got at least one solution (say R_k, defined above) but the entire system has no (global) solution.

As in (Pedrycz 1990), a search for a best approximate solution in order to optimize a performance index Q was used (Q – stands for the averaged *Hamming-fuzzy distance* between the desired and the obtained output fuzzy sets):

$$Q = (1/K)^* \sum_{k=1}^{k} |Y_k - R'{}^\circ X_k| ,$$

where R' is the *approximate solution*.

The search for the above mentioned task was started from a set of solutions of the individual equations. Their frequencies of appearance were used for trying to find a better approximate (global) solution for the entire system. A *GA* based searching method was applied for this goal, namely the *GA* described bellow, to obtain one column of the matrix representing the approximate (global) solution R. The same GA is repeated for each column to find the whole matrix R.

The chromosome was encoded in form of a $(n \times k)$ binary matrix for the following reasons: let $R_k = X_k \, a, Y_k$ be the greatest solution of the "k" equation as considered. For each pair (x_i, y_j) from the Cartesian space $X \times Y$ we obtain a string $\{a^1{}_{ji}, \ldots, a^K{}_{ji}\}$, where

$$a^k_{ji} = X_k(x_i)a, Y_k(y_j), k = 1, K .$$

It may occur (in usual cases) that some of $\{a^k{}_{ji}, k = 1, K\}$ will not be distinct, so the string is arranged so that the first $K'{}_{ij}$ elements do not repeat, and form a shorter string
$A_{ij} = \{a^1{}_{ij}, \ldots, a^{K'}{}_{ij}\}$, with their associate appearance frequencies $P_{ij} = \{p^1{}_{ij}, \ldots, p^{K'}{}_{ij}\}, K'{}_{ij} < K$.

For a fixed the "j" column of the matrix R is determined.

The "i" line of the chromosome's matrix will represent the A_{ij} vector in computing the fitness function. The evaluation of a chromosome is as follows: let

$$C = (c_{ik}, i = 1, \ n, \text{ for } k = 1, K\}c_{ik} \in \{0, 1\}$$

be a matrix representing a chromosome in the population, at a certain iteration. The fitness function is calculated as follows: let

$R'(x_i, y_j)$ be the average between $\{a^1{}_{ji}, \ldots, a^{K'}{}_{ji}\}$,
weighted with the weights $\{p^k_{ji} {}^* c_{ik}, k = 1, K'{}_{ij}\}$, for $i = 1, n$ namely

$$R'(x_i, y_j) = \sum_{k=1}^{K'ij} (p^k{}_{ji} {}^* c_{ik} {}^* a^k{}_{ij}) \bigg/ \sum_{k=1}^{K'ij} (p^k{}_{ji} {}^* c_{ik})$$

For this chromosome the fitness function is given by the following formula:

$$Q_j(C) = (1/K) * \sum_{k=1}^{K} |Y_k(y_j) - max_{i=1,n}(min\{X_k(x_i), R'(x_i, y_j)\})|$$

The fact that "j" is fixed is to be underlined.

Let us to consider a numerical example by taking into account a set of pairs of fuzzy sets as (X_k, Y_k) for $k = 1, 6 \ (K = 6)$:

$$X_k : \{x_1, x_2, x_3, x_4\} \Rightarrow [0, 1] \ (n = 4)$$

$$Y_k : \{x_1, x_2, x_3, x_4\} \Rightarrow [0, 1] \ (m = 4)$$

$$X = \begin{bmatrix} 1.0 & 0.4 & 0.3 & 0.1 \\ 0.4 & 1.0 & 0.6 & 0.4 \\ 0.5 & 0.5 & 0.8 & 0.9 \\ 0.6 & 0.4 & 0.7 & 1.0 \\ 0.9 & 0.6 & 0.4 & 0.3 \\ 0.5 & 0.6 & 0.8 & 1.0 \end{bmatrix} \quad Y = \begin{bmatrix} 0.5 & 0.5 & 0.7 & 1.0 \\ 0.3 & 0.3 & 1.0 & 0.9 \\ 0.4 & 0.2 & 0.2 & 0.6 \\ 0.5 & 0.6 & 1.0 & 0.3 \\ 0.4 & 0.3 & 0.5 & 0.7 \\ 1.0 & 0.6 & 0.3 & 0.2 \end{bmatrix}$$

The system of induced equations $(X_k{}^\circ R = Y_k, k = 1,6)$ is considered as follows: each *equation* treated separately has a solution, but the whole system has no a (global) one.

A classical algorithm for approximating the solution – as the one presented in (Pedrycz 1990) – led to the fuzzy relation as a result:

$$R = \begin{bmatrix} 0.4 & 0.3 & 0.7 & 0.7 \\ 0.4 & 0.3 & 1.0 & 0.9 \\ 0.3 & 0.3 & 1.0 & 0.6 \\ 0.4 & 0.6 & 1.0 & 0.6 \end{bmatrix}$$

and the performance index was accordingly calculated like an averaged Hamming distance between fuzzy sets:

$$Q = \sum_{j=1}^{4}(Q_j) = 0.15 + 0.12 + 0.26 + 0.17 = 0.70.$$

The GA proposed in (Negoita et al. 1994c) is performed for each column of the matrix R, which led to the following matrix after a relative small number of iterations ($\cong 10^3$, while the cardinal number of the searching space is $\cong 10^7$)

$$R = \begin{bmatrix} 0.40 & 0.37 & 0.70 & 1.00 \\ 0.37 & 0.30 & 1.00 & 0.90 \\ 0.35 & 0.00 & 0.25 & 0.37 \\ 0.50 & 0.60 & 0.25 & 0.60 \end{bmatrix}$$

with the performance index

$$Q = \sum_{j=1}^{4}(Q_j) = 0.13 + 0.12 + 0.20 + 0.15 = 0.60$$

This HIS method of solving fuzzy relational equations by GA has two advantages:

- the efficient use of the computer capabilities conferred by GA
- a better quantization of the solution space (by averaging $\{\alpha_{ji}^1, \ldots, \alpha_{ji}^K\}$, with certain weights, resulting from a weaker restriction against the method proposed in (Sanchez 1970). This is restricted because of the fact that the solution is chosen as the minimum of $\{\alpha_{ji}^1, \ldots, \alpha_{ji}^K\}$
- and against the method proposed in (Pedrycz 1990). This is restricted by the use of a probabilistic threshold between the same $\{\alpha_{ji}^1, \ldots, \alpha_{ji}^K\}$.

Further remarks and developments of this GA based HIS applied to solving fuzzy relational equations may be found in (Giuclea and Agapie 1996), namely with regards to real coded of chromosomes and to the use of a flexible fuzzy inference in form of the composed t-norm as from (Giuclea et al. 1996). In this extension, the *GA* will supplementary simultaneously optimize the composed t-norm and this is due to a better encoding of the chromosomes by including a separate gene for γ, the flexibility parameter of fuzzy inference.

An application of this immediately efficient *GA* based *HIS* method could be the *NRFM* (**N**ew **F**uzzy **R**easoning **F**uzzy **C**ontrollers) where the fuzzy relation matrix plays an important role regarding the performance of controllers (Park et al. 1994).

8.1.4 GA Based HIS for Prototype Based Fuzzy Clustering

One of the most important directions in application framework of fuzzy set theory to algorithms are the clustering methods (Zimmermann 1991). The usual optimization methodology of the clustering criteria provides alternative optimizations between the variables representing fuzzy memberships of the data to different clusters and the prototype variables (the variables determining the "geometry" of the clusters). But *GA* based *HIS* clustering has an interesting potential as a new clustering tool and this is especially due the low number of generations required to find good prototype parameters (Bezdek and Hathaway 1994). A *GA* based *HIS* fuzzy clustering in the above mentioned sense may be applied in case of the fuzzy c-means algorithm, which was developed by Bezdek as a generalization of the classical c-means algorithm.

Some difficulties during *GA* based *HIS* clustering with regard to preprocessing of the potential solutions appear namely in the step of discretizing and encoding the data. *GA* are proving to be useful in clustering if the data set encoding the fuzzy partition in form of chromosomes is suitable made.

But a disadvantage regarding the use of *GA* in clustering analysis is not to be neglected. A clustering application must solve the two task-problem of finding a fuzzy partition of input data and the prototypes of the discovered class. That asks for both the fuzzy partition and its representation to be encoded by a chromosome, in form of a linear string. But the genetic operators acting on such linear strings may unfortunately led to a loosing of an important amount of intrinsic information contained in the fuzzy partition. Much more, the chromosomes encoding in the form of linear string is not suitable to simultaneously represent both the fuzzy partitions and the prototypes of the discovered classes.

A *bidimensional encoding* of the chromosomes in form of a *matrix* seems to be more adequate for fuzzy clustering tasks as from (Dumitrescu et al. 1997), because it is an encoding manner achieving a natural encoding of the fuzzy partition or of its representation. This bidimensional encoding has the quality of keeping many relationship points of the data set. This reason was the central motivation for using the matrix encoding of the chromosomes for applications regarding knowledge handling in fuzzy expert systems (Negoita et al. 1994b). Both the fuzzy partitions as well as the corresponding prototypes are compactly represented by this multidimensional chromosome. The

experimental results in (Dumitrescu et al. 1997) confirm that a *GA* for clustering converges faster because it uses a multidimensional (matrix) encoding
instead of a linear one. In the same work some changes in the operators
used are presented as required by the matrix encoding of a chromosome. A
set of three crossover operators relying on matrix sub-blocks was developed
accordingly. The crossover performed on sub-matrices leads to a diminished
probability of less powerful offsprings appearing from two high performance
parents and many of the fuzzy sub-relation in the parents matrixes survive
to the generated offsprings, their destroying being much attenuated than in
case of (linear) string chromosomes crossover.

8.1.5 GA Based HIS in Fuzzy Modeling
and Systems Identification

A *GA* based *HIS* method for discrete time fuzzy model identification is proposed in (Negoita et al. 1995). A three-level optimization procedure is used
to minimize a quadratic performance index.

The discrete time dynamic system was considered in the form of a multi
input/single output one, with N inputs. The system was modeled as a system
of fuzzy relational equations. A trapezoidal shape for the membership function was used both for the fuzzy subsets defined on the universes of discourses
of fuzzy inputs and for the output. A special inference mechanism was used
in this model with the aim of improving the fuzzy inference procedure.

Design steps in *FS* modeling by starting from some measured input/output
data, are as follows: definition of the universes of discourse (domains and
ranges for input and output variables); specification of fuzzy membership
functions for the input and output variables; definition of fuzzy rules, building the fuzzy rules base; performing the numerical part of the inference algorithm (usually the *MAX-MIN* composition); defuzzification of fuzzy sub
sets through an inference mechanism applied to the fuzzy rule base (usually
represented by a relational matrix R). A usual supposition is that, in most
cases, the membership functions of the output have a constant form during
the optimization process.

But any classical method of fuzzy inference is relatively rigid as from
(Yag et al. 1994), where a flexible structure relies on *S-OWA* operators for
this mechanism was proposed, as below presented.

The $\wedge$ operator (typical *MIN* or *PROD*) is replaced by

$$\wedge' = (1 - \alpha)^* \wedge + \alpha^* arithmetic\ mean$$

But a *hybrid operator* was proposed in (Negoita et al. 1995), from the idea
that the *MIN* operator and at other times the *PROD* operator is sometimes
used for "$\wedge$" sometimes. This operator is replacing $\wedge$ in the first step of fuzzy
inference and looks as follows:

$$\wedge'' = (1 - \alpha)^* MIN + \alpha^* PROD$$

The above mentioned $\wedge''$ expression may be considered a logical generalization of the $\wedge$ operator because it is easy to remark that:

$$\text{for } \alpha = 1 \qquad \wedge'' = PROD$$

and

$$\text{for } \alpha = 0 \qquad \wedge'' = MIN$$

The $S\text{-}OWA\text{-}OR$ operator was proposed in (Yag et al. 1994) for the second step of the fuzzy inference:

$$\vee' = (1 - \beta)^* \vee + \beta^* \; arithmetic\; mean$$

But another hybrid operator was proposed in (Negoita et al. 1995), from the idea that the MAX operator is used for $''\vee''$ most of the time. This operator is replacing $\vee$ in the second step of fuzzy inference and looks as follows:

$$''\vee'' = (1 - \beta)^* PROD + \beta^* arithmetic\; mean$$

It must be noted that the classical form of the operator $PROD$ performing the second step of fuzzy inference mechanism is got in case of $\beta = 0$, namely $''\vee'' = PROD$.

The third step of the fuzzy inference mechanism (after truncation) is performing the outputs defuzzification, usually by the $BADD$ method (Tong 1978), consisting of:

$$y_0'(t) = \left(\Sigma_{j=1,\dots,M} b_j^\delta(t)^* y_j(t) \right) \Big/ \left(\Sigma_{j=1,\dots,M} b_j^\delta(t) \right),$$

where $b_j(t)$ are output (rule) truncation values, and $y_j(t)$ are output values.

The commonly used defuzzification method, *weighted sum method*, is got for $\delta = 1$.

The three above introduced parameters, α, β, δ are used in order to improve the fuzzy inference procedure. A GA based HIS optimization algorithm was built as a consequence for finding the optimal fuzzy relation R, the input fuzzy subsets and the optimal value for the coefficients α, β and δ. The chromosome encoding comprises three fields representing: the fuzzy relation, the parameters α, β, δ and the coefficients of the input membership functions.

The best values of α, β and δ parameters are got by the proposed GA acting on chromosomes in order to minimize the quadratic performance index:

$$J(I) = \Sigma_{j=1,\dots,1}(y_0'(t) - y(t))^2$$

where I is the number of examples and $y_0'(t)$ is computed by the above-mentioned formula, $y(t)$ being the crisp output value.

Using two different data sets as test examples proved the performance of this GA based algorithm. Using two slightly different adapted GA to each of the two applications experienced a good matching between GA structural parameters and the concrete database in each example.

A *simple GA* for the first example was used whose selection for the new generation is a competition between parents and children. Other structural parameters are as follows:

- 128 real genes; population size = 26; crossover probability = 80%; crossover type = *one point*;
- selection for crossover = *wheel roulette*; mutation probability = 5%.

A standard data set for system identification testing was used in this test example, namely the data set of Box Jenkins. This is a set of 296 pairs of input-output collected observation data for a fuzzy model of a gas furnace with two inputs and an output. Five triangular shaped fuzzy sub sets were used for each universe of discourse, but they were fixed membership functions for x_1^j, x_2^j, y_j.

The proposed simple GA performed both optimization of model fuzzy relational matrix and of the parameters α, β and δ in order to get a low order error $J(296)$. For $t_1 = 1, t_2 = 3$, where t_1, t_2, are the time delays, the following results were got:

$$J(296) = 0.380 \quad for \quad \alpha = 0.098, \ \beta = 0.478, \ \delta = 2$$

So the optimum is attained for $\alpha \neq 0, \beta \neq 0, \delta \neq 1$ that proves the applicability of the flexible inference. Much more, this result illustrates that $J(296)$ can be improved by using a simultaneous derivation of the fuzzy relation R and of the parameters, α, β and δ, in flexible inference as from Table 8.1. This table illustrates a comparison between the performance index $J(296)$ in case of different system identification algorithms.

Table 8.1. The comparative results on $J(296)$

Identification Method	ARMA (Box and Jenkins 1970)	Fuzzy (Yager et al. 1994)	Lee (Lee et al. 1994)	GA Based *HIS* (Negoita et al. 1995)
$J(296)$	0.71	0.469	0.407	0.380

A *special GA* for the second example was used whose selection for the new generation is an elitist one leading to expected good results in case of applications allowing deceptive problems. This selection for a new generation deletes 25% of the medium evaluated chromosomes, keeping 50% of the best individuals and 25% of the worst ones. Other structural parameters are as follows:

- 46 real genes; population size $= 32$; crossover probability $= 80\%$; mutation probability $= 5\%$.
- crossover type $=$ shuffle; the selection for crossover is based on exponential distribution
- the chromosome encoding fields are 3 (the first 30 genes are allocated to the fuzzy relational matrix R, 3 genes are for α, β, δ and 13 genes are for fuzzy membership functions)
- a random initialization was used for the parameters of the GA.

The characteristics of a *DC series motor*, as presented in (Park et al. 1994), were used in this test example as a data set for system identification testing. The input is the motor current $I \in \{1, 2, \ldots, 11\}$ and the output is the speed $N \in [0, 2000]$.

The fuzzy model proposed in this case (Park et al. 1994), is of the form:

$$Y(T) = X(t) \bullet R$$

where X correspond to I and Y to N the symbol "$\bullet$" denotes the fuzzy composition operator 6 fuzzy subsets for X and 5 fuzzy subsets for Y are considered.

The error performance index of data set as in (Park et al. 1994) is

$$J(10) = 7.61^*10^{-4}$$

for the coefficients α, β, δ as follows

$$\alpha = 0.936; \quad \beta = 0.695 \text{ and } \delta = 1 .$$

Table 8.2 illustrates a comparison between performance indexes $J(10)$ in case of different system identification algorithms.

Table 8.2. The comparative results on $J(10)$

The Method	$J(10)$
Optimization of relational matrix (Park 94 et al. 1994); random initialization based	1.370^*10^{-3}
Optimization of the membership functions based on a known relation [Park 94]	2.009^*10^{-4}
Special **GA** optimized system identification [Neg et al., 95d]; random initialization based	7.61^*10^{-4}

It is interesting to see that in these examples the hybrid operator for the inference is achieved for $\alpha \neq 0$ and $\alpha \neq 1$, this being a supplementary confirmation of the flexibility that features this kind of fuzzy inference.

An effective role of a fuzzy inference mechanism depends on some parameters, such as the fuzzy partition of the input/output universes of discourses that are traditionally decided in a subjective manner. This frame of subjectivity was favored by the traditional approach of building the fuzzy rule bases, by knowledge acquisition from human experts only.

A flexible *GA* based method for learning the parameters of a fuzzy inference system from examples must work by not involving subjectivity at all, such as proposed in (Fagarasan and Negoita 1995). The practice results indicate that a best performance of the fuzzy control system was obtained when *GA* searches the spaces of both the fuzzy membership function and the fuzzy relation matrix R. But a source of *subjectivity still remains* in deciding the number of fuzzy labels for both the input and output universes of discourse. A systematic design and optimization procedure of fuzzy inference systems is then needed. It was remarked that the performances of a fuzzy controller might be improved if the fuzzy reasoning model is supplemented by a genetic-based learning mechanism (Park et al. 1994).

Not only the performances of the fuzzy inference mechanism are improved by the *GA* based *HIS* method in (Fagarasan and Negoita 1995), but another consequence is a less complex structure for the system is achieved.

The task of the above mentioned method is to simultaneously learn the parameters of a fuzzy inference system, namely the number of fuzzy labels for inputs and the output, the shape and position of the membership functions and the structure of the rule base. And this new proposed method can be thought of as an alternative to other related methods in fuzzy modeling optimization, such as clustering techniques, template-based methods and combined *EC* with simulated annealing.

8.2 HIS Design Methods for High Performance GA

One of the most illustrative application examples useful to understand what hybridization of intelligent technologies in the framework of *HIS* means, such as personally treated by L. A. Zadeh, is described in (Lee and Takagi 1994), (Lee et al. 1995). A high performance *GA* structure was proposed in these two works.

The authors define a **D**ynamic **P**arameter **GA** (*DPGA*) as a *GA* whose dynamic parameters are controlled by a standalone fuzzy knowledge based system in a special manner. *DPGA* monitors the *GA* performance measures from an (external) evaluation system to control the most important *GA* parameters; namely the population size, mutation rate, and crossover rate. Different combination of either **GA** performance measures or current control settings are simultaneously the inputs of a fuzzy knowledge based system. Some of the *GA* parameters are the outputs of a fuzzy inference engine in the fuzzy knowledge based system.

The efficiency of *DPGA* in multiobjective optimization algorithms was proved by the authors with a method of integrated circuit layout generations, where the geometrical and timing specifications interact so that it is necessary to make some compromises along each of these directions to satisfy the circuits placement specifications (Lee et al. 1995).

An extension of this *DPGA* was made in (Lee and Esbensen 1997) in the form of a meta-level *EA* for automatically designing search control strategies for multiobjective *EA*. This meta-level technique is a designing approach of the fuzzy system used to monitor and control the search behavior. The evolved individuals represents some string-encoded fuzzy system parameters and the fitness function of meta-level *GA* evaluates how well the fuzzy system is able to control the *GA* search to maximize the search performance. This fitness function is a key to the success of such meta-level *GAs*. A suggested suitable solution for the fitness function is an offline one, maximizing the proposed measure of *GA* performance regarding the external environment and searching for a fuzzy system that maximizes this measure when the *GA* is run on several test functions. These test functions are strictly application specific, see for more details (Lee and Esbensen 1997).

8.3 GA Based Hybrid Intelligent Systems for Neural Networks Design and Optimization

One of the main technical purposes of this book is to convince the practitioners about the important fact that real-world applications have proved that intelligent techologies are complementary and not all as competitive.

A lot of applications proved in most cases the advantage of using these techniques in combination rather than exclusively standing alone, see detailed survey articles in such references as (Takagi 1996), (Negoita et al. 1994b), (Negoita et al. 1994a).

A typical example of intelligent techniques interacting in the *HIS* framework is given by the hybridization of *GA* with *NN* in applications.

This aggregation is made in three manners: *NN* (learning and configuration) based on *GA*, *NN* as tools for *GA* and cooperative systems (models and applications).

8.3.1 GA Based HIS Methods for NN Training

The impressive development of both *NN* theory and applications is in some extent shaded by the lack of a clear standard methodology for designing of an associated structure to any *NN* topology. A heuristically selected *NN* architecture was used as a consequence in most of the cases. But work during the past few years has emphasized *GA* and other *EC* techniques as an efficient and future promising tool in *NN* design.

NN structures may be designed by a *GA* combined with back propagation algorithms methodology, which varies the number of hidden layers and the processing elements (nodes) in a layer.

Another *GA* implication in *NN* design is for weights optimal adjustment (distribution) when the *NN* has a predetermined structure. A *NN* has usually a large number of weights in most applications, so a *GA* based *NN* optimization might be thought as one of the test problems for the performances of one *GA*.

Both the *NN* structures and weights can be directly encoded in form of binary strings, but the chromosome will have a large length in this coding. This is the reason for using a *GA* based *HIS* design either for the *NN* topology or for its weights.

The *EC* methods are also used in adaptive *NN* learning strategies to reward different learning function.

The work in previous years has emphasized the fact that *EP* is a more suitable *EC* method than *GA* is in *NN* building and this is because of a more efficient search in the space of *NN* architectures. The *EP* superiority on *GA* in *NN* building is natural and has a strong explanation in its essence from the *EC* elements point of view, namely: *EP* directly operates with *NNs*, not with their encoding, so a performance evaluation function is not needed; the crossover operator is not used, but mutation operator is the only offspring source in the parents space.

In this chapter only some aspects of *GA* implication in *NN* building are treated only.

The training process of a (supervised) *NN* is a minimization process of the mean square error between the desired outputs and the actual ones. Starting from this reason a *GA* was built in (Negoita and Mihaila 1995) for weights optimization of a simulated *STEPNET NN* trained to recognize handwritten digits. The main advantage of *STEPNET NN* is that its training algorithm simultaneously determines the topology of the network and the weights of the connection between neurons (Kerr et al. 1992). Two new *GA* operators were proposed in (Negoita and Mihaila 1995): *statistic guided mutation* and *statistic generation of chromosomes*, both of them relying on the construction of a statistic indicator of evolution, the statistic mask *Freq.(I)* of the frequencies of bit "1" for each generated position (gene) in the chromosomes among evolution.

8.3.2 HIS for NN Compiling into Fuzzy Rules Using GA

It is well known that *NNs* lack of human comprehensibility and therefore lack of any ability to explain make them to act as black boxes to users. A *GA* based HIS method as proposed in (Palade et al. 1998) is useful therefore for *NN* compiling (compiling of captured knowledge in the *NN* weight matrix) into a set of fuzzy rules (an equivalent *FS* to the *NN*). This proposed *HIS* could be used to develop a set of fuzzy rules from experimental data in the application area of control and pattern recognition, without any assistance from a human expert. The method improves *NN* based systems (see the neural controllers) conferring on them explanation facilities: a decision taken in a neural controller can be presented to the user in form of fuzzy rules as a consequence of this method, so an increased confidence in the controller action is achieved by this manner of modeling.

The *GA* is used to solve two distinct tasks of this *NN* compiling method:

– first, the *GA* is used for finding the best *FS* hierarchical structure (the fuzzy rules) (Palade et al. 1999), that is equivalent to the *NN*
– second, after the first task was achieved, the *GA* tunes the shapes and the number of linguistic terms (membership functions) of the hierarchical structure found at step 1

The fuzzy rules are extracted by an alternative to the *VIA* method (Thrun 94), the method of interval propagation. See more about this method in (Palade et al. 1999). This set of rules is used to obtain the initial population of solutions, giving a good starting point for the search for a suitable number and appropriate shapes of membership functions

The optimization goal of *GA* based *NN* compiling into a *FS* is to minimize the difference between the *NN* and its equivalent *FS* that wasachieved after the first two tasks are achieved

A step-by-step description of the *GA* used in (Palade et al. 1998) for mapping a *NN* into its equivalent *FS* looks as followings:

STEP 1 – equidistant partition of input space by suitably setting a number of certain membership functions, so that they divide the input space equally

STEP 2 – calculate – by *VIA* method (Palade et al. 1998) or better by method of interval propagation – (Palade et al. 1999) – the corresponding intervals of the output to the upper bases of the membership functions of the inputs

STEP 3 – an initial population of solutions is built by randomly initializeing the parameters of the lower bases of trapezoidal membership functions, based on the crisp rules got during *STEP 2*

STEP 4 – tune the shape of membership functions until the best fitness value becomes less than the target one, or the searching time reaches the pre- established number of generations. In the case of a stop criteria reaching the limited number of generations, go to *STEP 5*; stop otherwise because the minimal and optimal fuzzy model has been reached.

STEP 5 – let the mutation operator prune a membership function in one of the inputs or in the output; go to *STEP 2*

The chromosome evaluation is performed by a fitness function in the form of the sum of squared errors between the output value of the *NN* and the output value of the *FS*, with the parameters taken from the current chromosome in the population.

Some of the structural elements of *GA* based *NN* compiling into a *FS* are detailed as follows:

- *population size* was of 80–100 members
- the *chromosome* was 72 genes *real-encoded* (each real number being 0 and 1) in two fields: the first 7×4 numbers are the parameters of the trapezoidal membership functions of one input; the last 11×4 numbers represent the parameters of trapezoidal membership functions of the output. This representation of one input only along the length of the chromosome seems to be strange, but this is because the two inputs are interchangeable in the experiment. Four consecutive numbers compose a gene, which can be changed between two chromosomes.
- the crossover operator is a two-point one, one point is along the first 7 genes and the second point along the genes corresponds to the output. The crossover points are randomly chosen and must be divisible by 4.
- the mutation operator can alter any parameter of the gene. When the mutation operator alters the upper bases of a trapezoidal membership function of the input, automatically the upper bases of the corresponding membership functions are calculated (by *VIA* method or by interval propagation)
- the *solution* was found after around 300–400 generations.

But a main disadvantage of above-presented method is the exponential increasing of the fuzzy rules number with the increase of input space dimension (the fuzzy rules number is given by the product of membership functions number for each input). A second *GA* was applied in such circumstances to find an *optimal hierarchical structure of the FS* in order to reduce the number of membership functions and at the same time the number of fuzzy rules.

The building strategy of this hierarchical structure consists of collecting all the inputs having a closer relationship in the same fuzzy unit.

In a particular example, t the total number of fuzzy rules for a *NN* having 4 inputs, 1 output and 3 membership functions on each input, is $3 \times 3 \times 3 \times 3 = 81$ fuzzy rules. If there is an existing relationships between input 1 and input 2 for example, and also between inputs 3 and 4, these input are grouped in a hierarchical structure as in Fig. 8.2. Every fuzzy unit is described by 3×3 fuzzy rules, which mean an important diminishing of the total number of equivalent fuzzy rules.

A tree structure is built with 2^{n-2} node units having $1, 2, 4, 8, \ldots$ units on each level in the case of a *NN* that has n inputs, every *NN* input being assigned to a fuzzy unit in the tree. The fuzzy input is passed to the upper fuzzy unit in the tree if the fuzzy input has only one input.

An application was considered for a *NN* with inputs $n = 5$, the fuzzy units arrangement is as in Fig. 8.3a and an assigning of the tree units as in Fig. 8.3b.

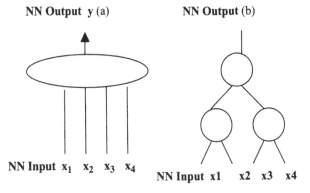

Fig. 8.2. An example of a black box **NN** (a) and its hierachical structure (b)

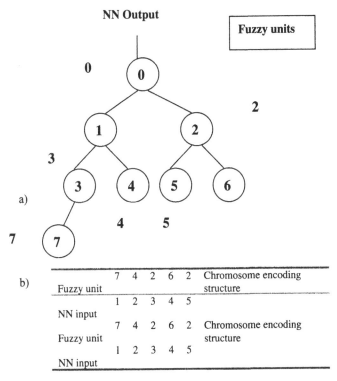

	7	4	2	6	2	Chromosome encoding
Fuzzy unit						structure
	1	2	3	4	5	
NN input						
	7	4	2	6	2	Chromosome encoding
Fuzzy unit						structure
	1	2	3	4	5	
NN input						

Fig. 8.3. (a) The fuzzy units tree corresponding to a **NN** with **n** inputs (b) The associated chromosome structure – the sequence of assigned (fuzzy unit) numbers to **NN** inputs

The developed *GA* for such a large *NN* is composed of two sub-procedures:

STEP 1 – find the best hierarchical structure of the equivalent *FS* to the *NN*

STEP 2 – tune the shapes and the number of membership functions of the hierarchical structure found during *STEP 1*

The chromosome is so encoded that each gene locus corresponds to a *NN* input and the gene allele is the number of the fuzzy unit which is assigned to that input – the same encoding method as in (Shimojima et al. 1995), see Fig. 8.3 for this encoding.

Other structural elements of the above-described *GA* are as follows:

– the crossover operator is of one point type
– the fitness function has the form:

$$f = E + a \cdot N \text{ , where}$$

E – is the sum of the squared error between the **NN** output and the output of the FS model

N – is the number of fuzzy rules and α is a coefficient.

The decoded corresponding *NN* hierarchical structure to example from Fig. 8.3 is as in Fig. 8.4:

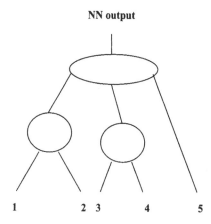

Fig. 8.4. Decoded corresponding NN hierarchical structure to coding in Fig. 8.3

8.3.3 Advanced Methods of GA Based HIS Design of NN

All kinds of *EA* are typically of fix length, either if they are *GA* optimizing a population of fixed length bit strings or if they are optimizing a fixed length n-dimensional real vector, as in case of *ES* and *EP*. But the fixed length structures, both the pre-established setting up of genotype length and meaning determination in advance for each gene, leads to little capacity ingenotype (structure) evolution.

The evolution of a problem (encoded) representation may be effective or not, depends strictly on the ability of its learning algorithm to modify the *structure of its own solutions*. By evolving its encoded structure, a variable length genotype may be able to learn not only the parameters of the solution, but also the main aspects of the encoding structure, namely: how many parameters there should be, what their meaning is and how they are interrelated. An evolutionary search operating with variable length structures introduces many degree of evolutionary search freedom that is missing in fixed length structures. In practice the application proved that the spectacular effects of changing their own structure are really relevant only if the variation of genotypic length structure is an integrant part of a much larger algorithmic dynamic with regard to iteration and selection operation. More details on this idea can be found in (Bahnzaf et al. 1998). The idea that unique spectacular properties of variable genotypic length structures appear in context of evolution only, is naturally rooted in genetics. *DNA* itself varies greatly in length and structure from species to species.

Complex organisms in nature present a kind of gene's specialization: structural genes, wqualitative, and adjusting genes, quantitative. The adjusting genes affect globally the structural genes. Another interesting evolution aspect is that a correspondence for a recombination operator that involves individuals with significant different genotypes is not possible to be found at any level of beings in the natural environment. These last two aspects of evolution in nature made possible the implementation of optimisation methods in the form of *VLGGA* (see Sect. 6.4) where the absolute position of the symbols usually do not provide information about the related feature. The *VLGGA* are successfully used in either achieving better performances for systems with complex structure or, at the same performances, a less complex structure of the system (Fagarasan and Negoita 1995). On the other hand, the gradient based learning algorithms for complex hybrid neuro-fuzzy architectures have a lot of local minima and a seriously time consumption complexity is involved as a consequence. These are the reasons forproposing a sophisticated *GA* learning based algorithm that solves simultaneously the optimization objectives of learning algorithms in fuzzy (algebraic) recurrent neural networks: both regarding the *FS-NN* performances (by its fuzzy weights matrix) and regarding the architecture (number of fully connected neurons) (see Arotaritei and Negoita 2002). A *GA* with variable length genotypes was developed in this work that offers a systematic way of getting a minimal neuro-fuzzy structure satisfying the above mentioned requested performance. This advantage is not to be neglected when a complex hybrid intelligent architecture must be designed without any previous details regarding it requested architecture.

8.4 GA Based Hybrid Intelligent Systems in Industrial Applications

The large variety of industrial applications of *GA* is impressive. For example *GA* application in different *CAD* branches could be included in the area of most representative industrial applications. Let us mention here just some applications in the electronics industry, namely *GA* applications in the main phases of digital *VLSI* layout design, partitioning, placement, routing and compaction. Some progress was made in analog circuits design too. Simple analog *VLSI* circuits are evolved by a parallel-implemented *GA* on a parallel computer system consisting of 20 distributed SPARC workstations in (Davis et al. 1994). But only some of the *GA* industrial or real world applications will be presented in this chapter.

8.4.1 GA Based Hybridization in Pattern Recognition and Image Processing

Pattern recognition and image processing is another application area of most *EC* techniques, *GA* especially. Different *EAs* are applied in this industrial area both in combination with other intelligent technologies and stand-alone. Subsection 8.4.1 describes just three applications of *GA* hybridization with other intelligent techniques in this industrial area.

A *GA* based *HIS* method was applied to protect a signature from being deliberately imitated (Yang et al. 1995). This *automatic signature verification system* functions *on-line* because the virtual strokes can be extracted and memorized in this manner by a computer, despite the fact that they are invisible for the user. The optimal features from these virtual strokes represent the authenticity index of the personal signature. The problem of predetermining the optimal feature combination needed in a signature verification system is very difficult and was solved by applying a special type of *GA*, called a *GA* with local improvement of the chromosomes or *Nagoya GA* (*NGA*). Both a *NGA* and a modified *NGA* variant will be described in Sect. 8.6.

The idea of *NGA* is to take over the microbial evolution and consists of a mutation evaluation in shorter intervals to improve the effectiveness of the mutation operator. The chromosome is evaluated applying a fuzzy network to handle the selected curve features. The intrapersonal variability of reference signatures is better absorbed by use of fuzzy membership functions for the evaluation.

A second *HIS* application presents a *GA* based method of minimizing the *NN* structure, with application to the recognition of US dollars (Takeda et al. 1994). This paper currency recognition method is based on paper currency masking.

The chromosome encodes the paper currency image. Firstly, some random masks are created based on a chromosomal encoding as follows: the mask position on the currency paper is a gene, 1 encoded if the correspondent region of the currency paper is masked, 0 encoded if the part is non-masked. A gradually optimization of the masks is achieved by a *GA* sequence of crossover, selection and mutation iteratively applied on the population of randomly created masks. Then these *GA* optimized masks become *NN* inputs. After learning is finished, the generalization of each *NN* is tested with masks using unknown data.

The recognition ability of the trained *NN* with optimized masks by *GA* proved to be superior to those trained *NN* with random masks generated with random numbers.

Another interesting application in pattern recognition is a *GA* based *HIS* approach to Chinese handwriting recognition (see Lin and Leou 1997).

The three main component modules of a Chinese handwriting recognition system are as follows: preprocessing module, feature extraction module and post-processing module. The types of Chinese handwriting distortions are: shape distortion, portion distortion and stroke distribution distortion. The recognition process steps run as follows:

- preprocessing stage – image processing operators are applied to input character images to make the input data suitable to be further processed
- *GA* stage – a *GA* is applied to determine the generalized normalization transform T containing 13 parameters for each input character
- post-processing stage – post-processing operators are used together with the generalized normalization transform obtained during *GA* stage to generate the corresponding normalized character

8.4.2 GA Based HIS in Different Forecasting Applications

Meteorological estimation of rainfall and temperature parameters is one of the most common real-world forecast application of *HIS*, relying mainly on *NN*s. But because of the large number of input variables large input vectors are used by these **NN**s. This drawback could possibly counteracted by *GA* based *HIS*: *GA*s are used there to reduce the input vector size with the supplementary precaution of keeping a high level of *NN* performance despite of this reduced size (McCullagh et al. 1997).

But the most requested real world forecast application involves the different branches of *financial engineering*, for example, applying modern time series forecasting approaches to the area of financial time series. The finance/banking engineering topic ranges over a high variety of applications. Most commonly, foreign currency exchange rate prediction uses *GA* based *HIS* methods for minimizing the prediction error for both: generating evolved bases of fuzzy rules, to cover many training examples as possible; and for the fine-adjustment of fuzzy membership functions of the fuzzy rule base.

GA-FS HIS are also efficiently used for intelligent financial decision making where *GA* are inducing fuzzy-rule bases in decision support systems for financial trading. This *HIS* method is a transparent one that is easily understood by the decision maker and is more advantageous over the machine learning methods that have difficulties in providing explicit explanations. In some applications, for example in credit evaluation, such kind of explanations are compulsory (Goonatlike and Campbell 1994).

8.4.3 GA Based HIS in Job Shop Scheduling

Aspects of *GA* based *HIS* applications to one of the well known resource allocation problem, namely job shop scheduling, are presented in (Lee et al. 1994). The resources to be allocated in job shop scheduling are *machines*, the basic tasks are *jobs* and each task consists of several sub-tasks called *operations*, which are correlated by precedence (sequencing) constraints. The schedules presented in the above mentioned paper must minimize the time required to achieve all jobs. That means to minimize the schedule length, under some constraints between jobs or between operations at the same machine. The developed *GA* for solving the above mentioned job shop scheduling problem make use of two special new proposed *GA* operators: precedence preserving crossover (*PPX*) and precedence preserving mutation (*PPS*). The applied *GA* only generates feasible schedules during the evolution and the method is applicable to job shop scheduling problems with different precedence constraints, either job precedence or operation precedence at the same machine.

A fuzzy mathematical programming model properly describes the aggregated production planning because the business environment in a manufacturing company is non-deterministic. A *GA* based *HIS* method is proposed for decision making in production planning (Wang and Fang 1997): a family of inexact solutions is found within an acceptable level instead of one exact optimal solution. A preferred solution is selected then by a decision maker from a convex combination of the above-mentioned family of inexact solutions. This extraction is interactively made by human-computer interaction after the *GA* is stopped. First, the decision maker is asked to deliver a requested degree of membership for the set of fuzzy optimal solution; then, the decision maker will chose a most preferred solution by an iterative devoted procedure.

The *GA* in this method uses an encoding with real numbers. The selection strategy is the roulette-wheel based proportional selection. A special mutation operator is used which moves along a weighted gradient direction to led the individuals to reach the fuzzy optimal solution set quickly. The points in dead areas of the solution space could be eliminated by the selection strategy.

Design of efficient distributed database (*DDS*) is a complex task of data allocation (Gen et al. 1997) solved by *GA* based *HIS* methodology. *DDS* led to significant advantages over centralized databases both in cost and response time. A *fuzzy DDS* design model is considered that allocates data to nodes, minimizing the sum of network communication cost for retrieval and update as well as local data storage costs. Triangular fuzzy numbers represents the various coefficients in DDS.

But the data allocation in *DDS* with fuzzy costs was transformed in a crisp problem according to the concepts of possibility and necessity, by setting the threshold value. Then a *GA* was applied to solve a data allocation problem in a *DDS* from the banking industry. The chromosome is a real encoded two-dimensional array. A uniform crossover is applied.

Some chromosomes may violate the problem constraints. These not feasible chromosomes are adjusted by a special repairing procedure during the *GA*. See (Gen et al. 1997) for more details on the repair algorithm.

8.5 GA Based Hybrid Intelligent Systems in Evolvable Hardware Implementation

This chapter is an introduction to an exciting and rapidly expanding industrial application area of *Evolutionary Computation (EC), especially of the Genetic Algorithms (GA): Evolvable Hardware (EHW).*

EHW has a central position in the large frame of *EC* applications because the hardware implementation of both genetic coding and artificial evolution methods led to a radical structural change of technology as a result. It means coming up with a new generation of machines. These machines evolve to attain a desired behavior, namely they have a *behavioral computational intelligence.* One of the main *EHW* implications in engineering design and automation is a conceptual/procedural one: no more difference exists between adaptation and design concerning these machines, these two concepts representing no longer opposite concepts (Negoita 2002a). *EHW* emergency shocked but rapidly produced an impressive concentration of the research forces of different countries in the world to effectively contribute in the implementation of real-world applications in *CI*.

8.5.1 EHW Definition and General Consideration – Implications on Engineering Design and Automation

A definition of *EHW* may be as follows: a sub-domain of artificial evolution represented by a design methodology (consortium of methods) involving the application of *EA* to the synthesis of digital and analog electronic circuits and systems (Kelly 1996). A more agreed definition among the practitioners might be: *EHW* is programmable hardware that can be evolved (Torresen 1997).

But some members of the scientific community acting in the area consider the term evolutionary circuit design more descriptive of *EHW* features. In additon another term used nowadays for the same work is *evolware* concerning to this evolvable ware with hardware implementation. This may well lead in the future to using the term *bioware which* concerns the possibility of evolving ware with biologic environments implementation. Even other environments are seen as possible evolvable media: *wetware* – real chemical compounds are to be used as building blocks or *nanotechnology* – relied on molecular scale engineering (Torresen 1997).

This new design methodology for electronic circuits and systems is not simply a fashion. It is suitable to special uncertain, imprecise or incomplete defined real-world problems, claiming a continuous adaptation and evolution. An increased efficiency of the methodology may be obtained by its application in the *HIS* framework that means in aggregation with other intelligent technologies such as *FS, NN* and *AIS*. The reason for using *EHW* in the above mentioned type of applications is its main advantage over the traditional engineering techniques for the electronic circuit design, namely the fact that the designer's job is very much simplified following an algorithm with a step sequence as below:

STEP 1 – problem specification
    1.1 – requirements specification of the circuit to be designed
    1.2 – specification of the basic (structural) elements of the circuit

STEP 2 – genome codification
    – an adequate (genotypic) encoding of the basic elements to properly
      achieve the structural circuit description

STEP 3 – fitness *calculation*
    – specification of the testing scheme used to calculate the genome
      fitness

STEP 4 – evolution (*automatically generation of the required circuit*)
    – generation of the desired circuit

The designer himself is involved by acting directly during the first three steps, while the fourth step is automatically generation of the circuit. The flow modality of both step 3 and step 4 leads to same categorizing classes criteria for *EHW*.

The idea of applying evolution to artificial systems has surfaced some decades ago, but the technology available at the time was not proper to implement this methodology in hardware.

Both computing technique development with increasing computational power and the appearance of programmable integrated circuits, especially their new generation – **Field Programmable Gate Arrays** (*FPGAs*) or most recently reconfigurable analogue arrays **Field Programmable Analogue Arrays** (*FPAAs*) and configurable digital chips at the functional block level (*open-architecture* **FPGAs**) make it possible for most companies to evolve circuits.

The second main *EHW* implication on engineering design and automation is the changing environment for electronics engineers, this profession being deeply changed: evolutionary approaches applied by *FPGAs* or open-architecture *FPGAs* to electronics makes hardware architectures as malleable as software, evolution doesn't care about complexity of how the evolving system works, the only limits to electronics design are the limits imposed by our own understanding.

8.5.2 EC Based Methods in EHW Implementation

This section is not at all focused on introducing all present methods that the scientists breaking new ground in the area of *EC* use to implement evolvable techniques and systems adaptivity in hardware. Also not all EHW applications are reviewed. A mention is to be made that some examples of *EHW* systems are applied in well known nowadays areas such as: analog and digital electronic circuits, cellular machines, controllers for autonomous mobile robots, pattern recognition and special *NNs* with dynamic topologies.

Biological models are applied in computer technique, adaptation and real-time learning as a natural consequence of moving from classical (hard) computing to $\mathbf{S}oft$ $\mathbf{C}omputing$ (SC). This allowed the passing from classical (*knowledge based AI*) to *behavioral computational intelligence* as was mentioned in the previous part of this chapter. The common framework of behavioral *AI* and of *SC* made possible the implementation of hardware circuitry with intrinsic logic elements specific to *EC*. This was used by real-world applications in those typical environments where human capacity of intervention is limited – nuclear plants, space applications, etc. – (Higuchi et al. 1994; Shimonara 1994; Negoita 1995; Negoita 1996).

8.5.3 EHW Architectures – The Ideal Environment for Implementing the Hardware Hybridization of Different Intelligent Techniques

Among the fundamental features of a living being, two are most essential for computer technique: *parallelism* and *co-operation of parts*. These are in fact the main features of the architectures used in hardware implementation of *EC* elements where the parallelism is "massive grain", that means a deeply developed one at different levels of the architecture. This chapter will refer to silicon *EHW*, but we must note that there is much research reported that is also focusing on wetware and nanotechnology *EHW* too.

The final aim of *EC* techniques development and their silicon implementation is to create architectures that approach an artificial brain, a computer having its own ability to reason and make decision, create emergent functionality and the possibility of self-creation and evaluation of its own structure.

The main types of hardware architectures with intrinsic *EC* logic elements are:

- embryological architectures
- emergent functionality architectures
- evolvable fault tolerant systems
- parallel evolvable architectures (of Higuchi type)

The specialized *FPGA* for embryological architectures is in essence a lattice substrate of *multiplexer cells* with associated *RAM*. This architecture is suitable for hardware implementation of both combinatorial and sequential complex logic circuits. Circuits are grown in silicon by architectures that are *RAM*-reprogrammable *FPGA* – a two-dimensional array consisting of dedicated (specialized cells) that can be interconnected by general interconnection resources (Marchal et al. 1996).

Emergent functionality architectures allow on-line evolution by real time development of a new functionality and new behavioral complexity for some autonomous agents typically in robotics or pattern recognition. Two main types of such architectures are known: co-operation architectures and subsumption architectures (Shimonara 1994).

An evolvable fault tolerant system was described in (Thompson 1995) that was evolved to control a real robot. It is an evolvable finite state machine held in *RAM*.

Parallel evolvable architectures (of Higuchi type) are a type of architecture with intrinsic *EC* elements – a real time reconfigurable hardware (**E**volvable **H**ard**W**are – *EHW*) as proposed by Higuchi (Higuchi et al. 1994; Higuchi 1996; Higuchi et al. 1996). In this paragraph of the book we will call this architecture an *EHW*. It has the greatest applicability nowadays, for example real time adaptivity of control systems in robotics or pattern recognition.

EHW is an ideal environment for implementing the hardware hybridization of different intelligent techniques (**F**uzzy **L**ogic – **FL**, **N**eural **N**etworks – **NN**, **G**enetic **A**lgorithms – **GA**) with **SC** learning methods (typically reinforcement learning or genetic learning), on a massive parallel architecture. This *HIS* technique confers very interesting and unique properties on *EHW* architectures: real time adaptation as a reaction to the work of the external environment (by real time modifying the architecture's structure) and robustness (slow decrease of the performances due to the environmental perturbations or hardware faults).

The concept of *EHW* architectures includes *three main component blocks* built in the same massive parallel manner – each block composed by the same number of specific sub-blocks (see Fig. 8.5, where just one of the n channels of sub-blocks was – for simplicity reasons – represented):

- the evolutionary component (*GA*) – a general structure, independent of application (a block of identical sub-blocks parallel implementing *GA*, with *GA* operators acting bit wise, that means bit by bit action in parallel, on different chromosomes); the *GA* block computes by genetic operations the architecture bits for *RLD* block.

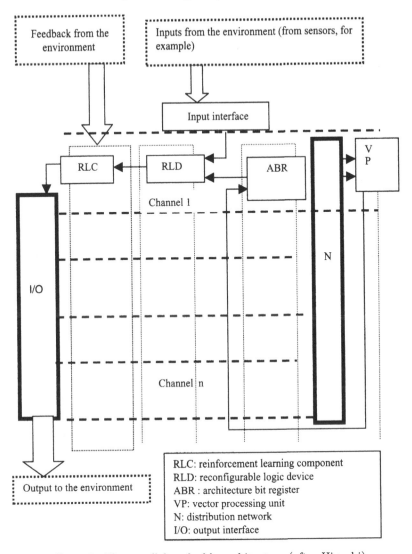

Fig. 8.5. The parallel evolvable architecture (after Higuchi)

- the hardware reconfigurable component (*RLD*) – a general *FPGA*-reliant structure, independent of application. This is a block of identical sub-blocks allowing real time change of configuration and functionality in hardware, as a response related to the behavior in the external environment. The *RLD* outputs depend on inputs and the architectural bits as provided from *GA* and the outputs of *RLD* are learned by the *RLC* component
- the learning component (*RLC*); this block has an application dependent architectural structure and usually is externally implemented. The *RLC*

computes the reinforcement signal (reward or penalty) as an environment reaction for adaptively modifingy the fitness function of GA chromosomes

The Higuchi EHW architecture typically performs learning of the table representing the relation between the inputs and outputs, a usual application both in complex control systems of robots and in pattern recognition.

Some problems arise from the hardware implementation of GA, especially of those devoted to EHW architectures because of the required compromise between complexity of .efficient GA and decreased architectural complexity that can solve the optimization of hardware devices (Mihaila et al. 1996). Good solutions of this problem call for further research a simplified hardware implementation of efficient, but complex GA such as *Nagoya GA* or even *VLGGA*.

8.5.4 A Suitable GA for Implementing the Hardware Hybridization of Intelligent Techniques in EHW Architectures

Real-time adaptation in EHW systems means changes in their architectures, as a reaction to the variable external environmental conditions and as a privilege conferred by the properties of the technological support of these circuits. This approach is a step further than the standard approaches so far, where adaptation only affects software structures.

RLD is a principal component part of the Higuchi EHW type of architectures. As previously mentioned, the general structure of the RLD block is based on the technological support of $FPGA$. RLD related simulations were made considering one of the most used $FPGA$, a PLD structure that is composed of a lot of identical macro-cells. A $GAL16V8\ PLD$ circuit was simulated (like those delivered by Lattice Semiconductors), see (Negoita et al. 1997; Negoita 1997a,b).

The tool selected for this application was a *GA with local improvement of chromosomes*. The main reason of local improvements in chromosomes is just the application of *genetic operators at the level of each group gene* in the chromosome. Namely, the chromosome is implemented as a group of segments, group genes, and a mutation operator is applied to each set of group genes. *GA with local improvement of chromosomes* has a step sequence as described in (Furuhasui et al. 1994; Yang et al. 1995).

Experimental results were obtained (Negoita 1997a; Negoita et al. 1997), as a comparison of the solutions found to this application problem by using both a classical Goldberg's GA (SGA) and a GA of Nagoya type (NGA) as thought in the two above mentioned works.

SGA has used a roulette-wheel selection and the new generation was achieved by a total replacement of the old generation. The improved NGA used in the work had $m = 5$ – the copies number for a group gene, and the number of group genes (segments) was $N = 8$, namely the same as the number of columns in the connection matrix of *GAL16V8* circuit. In contrast with the Nagoya algorithm used in (Furuhasui et al. 1994; Yang et al. 1995), after applying the Nagoya mutation on all chromosomes, *Step 3* of the improved *NGA* includes the mutation too, not the reproduction and crossover only, the parents selection type and the replacement of the population for the new generation being kept.

8.5.5 GA Based Hybrid Intelligent Systems and Emergent Technologies. Application to NN Based Multi-Classifier Systems for Gene Identification in DNA Sequences

A *GA-NN HIS* was presented in (Ranawana and Palade 2004) that performs the identification of Escherichia Coli (*E.Coli*) promoter sequences in strings of *DNA*. As each gene in *DNA* is preceded by a promoter sequence, the successful location of an E.Coli promoter leads to the identification of the *E.Coli* gene in the *DNA* sequence. A promoter is a region in the *DNA* where the transcription initiation takes place.

A *NN* based multi-classifier system, *MultiNNProm* (*Multiple Neural Network Based System for Promoter Recognition*), see Fig. 8.6, was namely used for the identification of these *promoter* sites. The proposed system contains four *NNs*, to which the same *DNA* sequence is presented using four different encoding methods. The outputs of the individual *NNs* are then passed

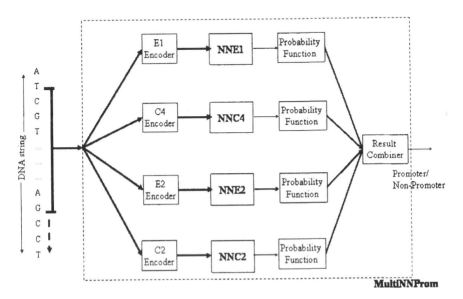

Fig. 8.6. The *GA-NN HIS* architecture of *MultiNNProm*

through a probability function and finally combined by an aggregation function. This aggregation function aggregates the results in order to provide an answer as to whether the presented sequence is an *E.Coli* promoter or not.

The combination (aggregation) function used in *MultiNNProm* combines the classification results of the individual classifiers by calculating the entropy of each output and then aggregating them through a weighted sum. The values of the aggregating weights were obtained by initializing a random population of weights. Then a GA runs in order to obtain an optimal set of weights, by using the classification accuracy of the combined classifier as fitness value. Genetic algorithms are able to search a large space of candidate hypotheses to identify the best hypothesis. They have been proven to be useful in situations where an evaluation method to calculate the fitness of a candidate hypothesis is available.

An overview of this *GA-NN HIS*, as from Fig. 8.6, shows that the system is a *NN* based multi-classifier system. Four *NNs* were developed, that are called *NNE1*, *NNE2*, *NNC2* and *NNC4*, into which the same *DNA* sequence is inputted using four different encoding methods: *E1*, *E2*, *C2* and *C4* respectively. The outputs of the individual *NNs* are then passed onto a probability builder function, which assigns probabilities as to whether the presented sequence is a E.Coli promoter or not. Finally, the outputs of the probability functions are aggregated by a result combiner, which combines the results and produces a final result as to whether the given sequence is an *E.Coli* promoter or not. The final output is of the "*yes*" or "*no*" form.

The combination of results obtained through multiple *NN* classifications was made by using an aggregation method called *LOP2*. This proposed method is a modification of the **L**ogarithmic **O**pinion **P**ool (*LOP*) (Hansen and Krogh 1995) – which proved to be more adequate and produce better results when using GA. More regarding the methods for combining classifiers and on *LOP2* especially are to be found in (Ranawana and Palade 2004).

References

Abdel R E, Elhadidy M A, "Modeling and forecasting the daily maximum temperature using abductive machine learning", Oceanographic Literature Review, vol. 43(1), pp. 25–37, 1996

Adamczak R, Duch W, Neural networks for structure-activity relationship problems, in Proceedings of the 5th International Conference on Neural Networks and Soft Computing, Zakopane, 2000: 669–674

Alexandru D, Combastel C, Gentil S (2000) Diagnostic decision using recurrent neural network. In: Proc. of the 4th IFAC Symposium on Fault Detection, Supervision and Safety for Technical Processes, pp. 410–415, Budapest

Arotaritei, D: Dynamic Adapted Gradient Algorithm for Hybrid Recurrent Structure. World Multiconference on Systemics, Cybernetics and Informatics ISAS/SCI 2001, Orlando, Florida, USA (2001), pp. 568–571

Arotaritei D, Negoita Gh M (2002) Optimisation of Recurrent NN by GA with Variable Length Genotype. In: McKay B, Slaney J (eds) AI2002: Advances in Artificial Intelligence. Springer-Verlag, Berlin Heidelberg New York, pp. 691–692

Avouris N M, "Cooperating Knowledge-Based Systems for Environmental Decision Support", Knowledge-Based Systems, vol. 8(1), pp. 39–54, 1995

Avouris N M, Kalapanidas E, "Expert Systems and Artificial Intelligence Techniques in Air Pollution Prediction", in Mathematical Modelling of Atmospheric Pollution, edited by Moussiopoulos N, Thessaloniki, 1997

Bahnzaf W, Nordin P, Keller E R, Francone D F (1998) Genetic Programming. An Introduction on Automatic Evolution of Computer Programs and Its Application. Morgan Kaufmann Publishers Inc.

Bajaj P, Keskar A (2003) Soft Computing-Based Design and Control for Vehicle Health Monitoring. In: Palade V, Howlett J R, Jain L (eds) Knowledge-Based Intelligent Information and Engineering Systems. Springer-Verlag, Berlin Heidelberg New York, Part I, pp. 563–569

Barbara Finlayson-Pitts and James Pitts N J, Atmospheric Chemistry: Fundamentals and Experimental Techniques, Wiley: New York, 1986

Baylor A (2000) Beyond Butlers: Intelligent Agents as Mentors. J Educational Research 22(4): 373–382

Benfenati E, Mazzatorta P, Neagu C-D, Gini G, Combining classifiers of pesticides toxicity through a neuro-fuzzy approach, 3rd International Workshop on Multiple Classifier Systems MCS2002, 24–26 June 2002, Cagliari, Italy, Lecture Notes in Computer Science LNCS 2364 (Roli F, Kittler J: eds.), 293–303, Springer Verlag Berlin Heidelberg

Benitez J M, Blanco A, Delgado M, Requena I., "Neural methods for obtaining fuzzy rules", Mathware Soft Computing, vol. 3, pp. 371–382, 1996

Benitez J M, Castro J L, Requena I, "Are Artificial Neural Networks Black Boxes?", IEEE Transactions on Neural Networks, vol. 8, no. 5, pp. 1157–1164, 1997

Bezdek J C, Hathaway R J (1994) Optimization of Fuzzy Clustering Criteria Using Genetic Algorithms. In: Proceedings of WCCI'94, September 24, pp. 589–594

Bloom B S (1984) The 2-sigma problem: the search for methods of group instruction as effective as one-to-one tutoring. Educational Researcher 13 (6): 4–16

Box G E P, Jenkins G M (1970) Time Series Analysis, Forecasting and Control. Holden Day, San Francisco, CA

Boznar M, "Pattern Selection Strategies for a Neural Network-Based Short Term Air Pollution Prediction Model", in Proc. of the IEEE International Conference on Intelligent Information Systems, IIS'97, p. 340–344, 1997

Bradley D W, Ortega-Sanchez C, Tyrrell A (2000) Embryonics + Immunotronics: A Bio-Inspired Approach to Fault Tolerance. In: Proceedings of the 2nd NASA/DoD Workshop on Evolvable Hardware, IEEE Computer Society, Los Alamitos, California, pp. 215–223

Bradley D W, Tyrrell A (2001) The Architecture for a Hardware Immune System. In: Proceedings of the 3rd NASA/DoD Workshop on Evolvable Hardware, IEEE Computer Society, Los Alamitos, California, pp. 193–200

Bridle J S, "Training stochastic model recognition algorithms as networks can lead to maximum mutual information estimation of parameters", in Advances in Neural Information Processing Systems, edited by Touretzky D J, Kaufmann, New York, vol. 2, pp. 211–217, 1990

Brown M, Harris C J (1994) Neuro-fuzzy adaptive Modelling and Control. Prentice-Hall

Buckley J J, Hayashi Y, Czogala E (1993) On the equivalence of neural nets and fuzzy expert systems, Fuzzy Sets and Systems 53/2, 129–134

Buckley J J and Hayashi Y, "Neural nets for fuzzy systems", Fuzzy Sets and Systems, vol. 71, pp. 265–276, 1995

Buiu C (1999) Artificial intelligence in education – the state of the art and perspectives (ZIFF Papiere 111). Germany, Fern University, Institute for Research into Distance Education (ERIC Document Reproduction Service No. ED 434903)

Calado J M F, Sa da Costa J M G (1999) An expert system coupled with a hier archical structure of fuzzy neural networks for fault diagnosis. J Applied Mathematics and Computer Science 3 (9): 667–688

Castellano G, Castiello C, Fanelli A M, Mencar C (2003) Discovering Prediction Rules by a Neuro-fuzzy Modeling Framework In: Palade V, Howlett J R, Jain L (eds) Knowledge-Based Intelligent Information and Engineering Systems. Springer-Verlag, Berlin Heidelberg New York, Part I, pp. 1243–1248

Chen W (2003) Supporting Collaborative Knowledge Building with Intelligent Agents. In: Palade V, Howlett J R, Jain L (eds) Knowledge-Based Intelligent Information and Engineering Systems. Springer-Verlag, Berlin Heidelberg New York, Part II, pp. 238–244

Chimmanee S, Wipusitwarakun K, Runggeratigul S (2003) Adaptive Per-application Load Balancing with Neuron-Fuzzy to Support Quality of Service for Voice over IP in the Internet. In: Palade V, Howlett J R, Jain L (eds) Knowledge-Based Intelligent Information and Engineering Systems. Springer-Verlag, Berlin Heidelberg New York, Part I, pp. 533–541

Coello C A (1999) A Survey of Constraint Handling Techniques used with Evolutionary Algorithms. Lania-RI-99-04, Laboratorio Nacional de Informática Avanzada

Cordon O, Herrera F, Lozano P [1997] On the Combination of Fuzzy Logic and Evolutionary Computation: A Short Review and Bibliography, in Pedrycz W, (Ed.), *Fuzzy Evolutionary Computation*, Kluwer Academic Publisher Boston, London, Dodrecht, pp. 41–42

Cormen T H, Leiserson C E, Rivest R L (1990) Introduction to Algorithms. The MIT Press

Craciun M, Neagu C-D, Using Unsupervised and Supervised Artificial Neural Networks to study Aquatic Toxicity, Proceedings of the 7th International Conference on Knowledge-Based Intelligent Information & Engineering Systems – KES2003, Oxford – UK, 3–5 Sept. 2003, 901–908, LNAI Springer-Verlag

Cronin M T D, Aptula A O, Duffy J C, Netzeva T I, Rowe P H, Valkova I V, Schultz T W (2002) Comparative assessment of methods to develop QSARs for the prediction of the toxicity of phenols to Tetrahymena pyriformis. Chemosphere 49: 1201–1221

Cronin M T D, Schultz T W, Pitfalls in QSAR. Journal of Molecular Structure (Theochem), 2003; 622: 39–51

Cronin M T D, Schultz T W, Structure-toxicity relationships for phenols to Tetrahymena pyriformis. Chemosphere, 1996; 32: 1453–1468

Cutello V, Nicosia G (2003) Noisy Channel and Reaction-Diffusion Systems: Models for Artificial Immune Systems. In: Palade V, Howlett J R, Jain L (eds) Knowledge-Based Intelligent Information and Engineering Systems. Springer-Verlag, Berlin Heidelberg New York, Part II, pp. 496–502

Cutello V, Nicosia G, Pavone M (2003) A Hybrid Immune Algorithm with Information Gain for the Graph Coloring Problem. In: Cantu-Paz E, Foster J, Kalaynmoy D, Lawrence D D, Rajkumar R, O'Reilly U-M, Beyer H-G, Standish R, Kendall G, Wilson S, Harman M, Wegner J, Dasgupta D, Potter M, Schultz A, Dowsland K, Jonoska N, Miller J (eds) Genetic and Evolutionary Computation – GECCO 2003, Springer Verlag Berlin Heidelberg New York, Part I, pp. 171–182

da Rocha A F, Neural Nets: A Theory for Brains and Machines, Lecture Notes in Artificial Intelligence, Springer-Verlag, 1992

da Rocha A F, "The fuzzy neuron: biology and mathematics", in Proc. of IFSA'91 Brussels, vol. 1, pp. 176–179, 1991

Dasgupta D, Yu S, Majmudar NS (2003) MILA – Multilevel Immune Learning Algorithm. In: Cantu-Paz E, Foster J, Kalaynmoy D, Lawrence D D, Rajkumar R, O'Reilly U-M, Beyer H-G, Standish R, Kendall G, Wilson S, Harman M, Wegner J, Dasgupta D, Potter M, Schultz A, Dowsland K, Jonoska N, Miller J (eds) Genetic and Evolutionary Computation – GECCO 2003, Springer Verlag Berlin Heidelberg New York, Part I, pp. 183–194

Davis M, Liu L, Elias G J, [1994] VLSI Circuit Synthesis Using a Parallel Genetic Algorithm in *Proceedings of ICEC'94*, vol. 2, pp. 104–109

de Castro L N, Timmis J I (2002) Artificial Immune Systems: A New Computational Approach. Springer Verlag, London

de Castro L N, von Zuben F J (2000) The Clonal Selection Algorithm with Engineering Application. In: Proceedings of the Genetic and Evolutionary Computation Conference, Workshop on Artificial Immune Systems and Their Applications, pp. 36–37

Deaton R, Garzon M (2001) Fuzzy Logic with Biomolecules, In: J Soft Computing 5: 2–9, Springer Verlag

Deaton R, Garzon M, Rose J A, Murphy R C, Stevens Jr S E, Franceschetti D R (1997) A DNA Based Artificial Immune System for Self-Nonself Discrimination, In: J Proceedings of The IEEE System, Man, and Cybernetics I: 862–866

Duin R P W, Tax D M J,: Experiments with Classifier Combining Rules. Lecture Notes in Computer Science, Springer-Verlag, Berlin, 2000; 1857: 16–29

Dumitrescu D, Stan I, Dumitrescu A (1997) Genetic Algorithms in Fuzzy Clustering-Multidimensional Encoding. In: Proceedings of EUFIT'97, Aachen, Germany, pp. 705–708

Enbutsu I, Baba K, Hara N, "Fuzzy Rule Extraction from a Multilayered Network", in Proceedings of IJCNN'91, Seattle, 1991

Exner T E, Brickmann J (1997) New Docking Algorithm Based on Fuzzy Set Theory, Journal Molecular Modelling, 3: 321–324

Fagarasan F, Negoita Gh M (1995) A Genetic-Based Method for Learning the Parameter of a Fuzzy Inference System. In: Kasabov N, Coghill G (eds)

Artificial Neural Networks and Expert Systems, IEEE Computer Society Press, Los Alamitos California, pp. 223–226

Finin T, Labrou Y, Mayfield J (1998), KQML as an Agent Communication Language, in Bradshaw J M, (ed), Software Agents, MIT Press, Cambridge MA

Forrest S, Javornik B, Smith R E, Perelson A S (1993) Using Genetic Algorithms to Explore Pattern Recognition in the Immune System. J Evolutionary Computation 1(3): 191–211

Forrest S, Perelson A, Allen L, Cherukuri R (1994) Self-Nonself Discrimination in a Computer. In: Proceedings of the IEEE Symposium on Research in Security and Privacy, pp. 202–212

Fuller R, Introduction to Neuro-Fuzzy Systems, Advances in Soft Computing Series, Springer-Verlag: Berlin, 1999

Furuhashi T, Nakaoka K, Uchikawa Y (1994) A New Approach to Genetic Based Machine Learning and an Efficient Finding of Fuzzy Rules. In: Proceedings of The IEEE/Nagoya University WWW on Fuzzy Logic and Neural Networks/Genetic Algorithms. Nagoya, Japan, pp. 114–122

Garg R, Kurup A, and Hansch C, Comparative QSAR: on the toxicology of the phenolic OH moiety, *Crit. Rev. Toxicol.*, 2001; 31: 223–245

Gauvin D, (1995), Un environnement de programmation oriente agent, 3eme journees francophones sur l'IA distribuee et les systemes multiagents, St-Baldoph, Savoie, France

Gen M, Tsujimura Y, Ko S (1997) Allocation Strategy for Distributed Data Base System with Fuzzy Data Using Genetic Algorithms. In: Proceedings of EUFIT'97. Aachen, Germany, pp. 737–742

Gini G, Benfenati E, Neagu C-D, Training through European Research Training Networks – Analysis of IMAGETOX, 3rd IEEE International Conference on Information Technology Based Higher Education and Training ITHET 2002, July 4–6, 2002, Budapest

Gini G, Lorenzini M, Benfenati E, Brambilla R, Malvé L: Mixing a Symbolic and a Subsymbolic Expert to Improve Carcinogenicity Prediction of Aromatic Compounds. Kittler J, Roli F eds: Multiple Classifier Systems., 2001; 126–135, Springler-Verlag, Berlin

Gini G, Predictive Toxicology of Chemicals: Experiences and Impact of AI Tools, AI Magazine, AAAI Press, 2000; 21/2: 81–84

Giuclea M, Agapie A (1996) A GA Method for Approximate Solving Fuzzy Relational Equations. In: Reusch B, Dascalu D (eds), Real-World Applications of Intelligent Technologies (part II), pp. 146–151

Giuclea M, Agapie A, Fagarasan F (1996) A GA Approach for Closed-Loop System Identification. In: Proceedings of EUFIT'96, Aachen, Germany, pp. 410–416

Gonzales F, Dasgupta D, Gomez J (2003) The Effect of Binary Matching Rules in Negative Selection. In: Cantu-Paz E, Foster J, Kalaynmoy D, Lawrence D D, Rajkumar R, O'Reilly U-M, Beyer H-G, Standish R,

Kendall G, Wilson S, Harman M, Wegner J, Dasgupta D, Potter M, Schultz A, Dowsland K, Jonoska N, Miller J (eds) Genetic and Evolutionary Computation - GECCO 2003, Springer Verlag Berlin Heidelberg New York, Part I, pp. 195–206

Goonatilake S, Campbell J A (1994) In: Proceedings of IEEE/Nagoya University WWW Conference on Fuzzy Logic and Neural Networks/Genetic Algorithms, August, Nagoya, Japan, pp. 143–155

Goonatilake S, Treleaven P (1996) Intelligent Systems for Finance and Business. John Wiley & Sons, Chichester New York Brisbane Toronto Singapore

Gori M, and Scarselli F: Are multilayer perceptrons adequate for pattern recognition and verification? IEEE Transactions on Pattern Analysis and Machine Intelligence 1998; 20/11: 1121–1131

Gustafson D E, Kessel W C (1979) Fuzzy clustering with a fuzzy covariance matrix, In: *Proc IEEE Conf Decision Contr.* San Diego, CA, pp. 761–766

Hagan M, Beale M, Neural Network Design, PWS: Boston, 1996

Hansch C, Hoekman D, Leo A, Zhang L, Li P: The expanding role of quantitative structure-activity relationship (QSAR) in toxicology. Toxicology Letters, (1995; 79: 45–53

Hansen J V, Krogh A (1995) A General Method for Combining in Predictors Tested on Protein Secondary Structure Prediction, citeseer.ist.psu.edu/324992.html

Hart E, Ross P (1999) The Evolution and Analysis of a Potential Antibody Library for Use in Job-Shop Scheduling. In: Corne D, Dorigo M, Glover F (eds) New Ideas in Optimization, McGraw Hill London, pp. 195–202

Harvey I (1991) Species Adaptation Genetic Algorithms: a basis for a continuing SAGA. In: Proceedings of the First European Conference on Artificial Life. Toward a Practice of Autonomous Systems, MIT Press, Cambridge, York, pp. 346–354

Harvey I (1992) The SAGA Cross: The mechanism of Recombination for Species with Variable-Length Genotypes. J Parallel Problem Solving from Nature 2. Elsevier Science Publisher B V

Hashem S, "Optimal Linear Combinations of Neural Networks", Neural Networks, vol. 10(4), pp. 599–614, 1997

Hatzilygeroudis I and Prenzas J, Constructed Modular Hybrid Rule Bases for Expert Systems, Int. Journal on Artificial Intelligence Tools, 2001; 10: 87–105

Haykin, S, Neural Networks: A Comprehensive Foundation, IEEE Press: New Jersey., 1994

Hecht-Neilson R, Neurocomputing, 1990, Addison-Wesley Pub Co

Helma C, Gottmann E, Kramer S,: Knowledge discovery and data mining in toxicology. Statistical methods in medical research, 2000; 9: 131–135

Herrera F, Lozano M, Verdegay I L (1993) Tunning Fuzzy Logic Controllers by Genetic Algorithms. *Technical Report # DECSAI – 93102*, June, Universidad de Granada, Espagna

Higuchi T (1996) Evolvable Hardware with Genetic Learning. In: Proceedings of IEEE International Symposium on Circuits and Systems, ISCAS96. May 13, Atlanta, USA

Higuchi T, Iba H, Manderick B (1994) Evolvable Hardware. In: Kitano H, Hendler J (eds) Massively Parallel Artificial Intelligence, AAAI Press/MIT Press, pp. 339–420

Higuchi T, Iwata M, Kajtani I, Iba H, Hirao Y, Furuya T, Manderick, B (1996) Evolvable Hardware and Its Applications to Pattern Recognition and Fault-Tolerant Systems. In: Sanchez E, Tomassini M (eds) Towards Evolvable Hardware. Springer Verlag, Heidelberg, pp. 118–136

Hilario M, "An Overview of Strategies for Neurosymbolic Integration", chap. 2 in Connectionist-Symbolic Integration: From Unified to Hybrid Approaches, edited by Sun R, and Alexandre F, Lawrence Erlbaum Associates, 1997

Hunt J E, Cooke D E, Holstein H (1995) Case Memory and Retrieval Based on the Immune System. In: Weloso M, Aamodt A (eds) Case Based Reasoning Research and Development, LNAI 1010, Springer Verlag, pp. 205–216

Ishida Y (2004) Immunity-Based Systems: A Design Perspective. Springer Verlag, Berlin, Heidelberg New York (ISBN: 3-540-00896-9)

Jacobs R A, Jordan M I, Barto A G, "Task decomposition through competition in a modular connectionist architecture: the what and where vision tasks", Cognitive Science, vol. 15, pp. 219–250, 1991

Jacobs R A, Methods for combining experts' probability assessments. Neural Computation 1995; 7/5: 867–888

Jagielska I, "Linguistic Rule Extraction from Neural Networks for Descriptive Datamining", in Proc. of 1998 Second International Conference on Knowledge-Based Intelligent Electronic Systems, edited by Jain L C, Jain R K, Adelaide, 1998

Jagielska I, Matthews C, Whitfort T, "A Study in Experimental Evaluation of Neural Network and Genetic Algorithm Techniques for Knowledge Acquisition in Fuzzy Classification Systems", in Proc. of IEEE ICNN Conf. on Neural Networks, Houston, 1997

Jagielska I, Matthews C, Whitfort T (1999) An investigation into the application of neural networks, fuzzy logic, genetic algorithms, and rough sets to automated knowledge acquisition for classification problems, Neurocomputing 24, 37–54

Jang J S, Sun R (1995) Neuro-fuzzy modeling and control. J Proc IEEE 83(3): 378–405

Jang R, ANFIS J-S : Adaptive-network-based fuzzy inference system. IEEE Transactions on Systems, Man, and Cybernetics, 1993; 23: 665–685

Jang R J S, Sun C T, "Functional equivalence between radial basis function networks and fuzzy inference systems", IEEE Transactions on Neural Networks, vol. 4(1), pp. 156–159, 1993

Jun J -H, Lee D -W, Sim K-B (1999) Realization of Cooperative and Swarm Behaviour in Distributed Autonomous Robotic Systems Using Artificial Immune System. In: Proceedings of the IEEE System, Man and Cybernetics Conference, Part 4, pp. 614–619

Kandel A, Teodorescu H N, Arotaritei D: Analytic Fuzzy RBF Neural Network. Proceedings of the 17th Annual Meeting of the North American Fuzzy Information Processing Society NAFIPS'98, Pensacola Beach, Florida, USA (1998)

Kasabov N, "Learning Fuzzy Rules and Approximate Reasoning in Fuzzy Neural Networks and Hybrid Systems", Fuzzy Sets and Systems, vol. 82, pp. 135–149, 1996

Katritzky A R, Lobanov V S, and Karelson M: CODESSA Comprehensive Descriptors for Structural and Statistical Analysis. Reference Manual ver. 2.0, Gainesville (1994)

Kelly M (1996) Fit for Future? Evolutionary Computation in Industry. J EvoNews 2: 1–3

Kelsey J, Timmis J I (2003) Immune Inspired Somatic Contiguous Hypermutation for Function Optimization. In: Cantu-Paz E, Foster J, Kalaynmoy D, Lawrence D D, Rajkumar R, O'Reilly U-M, Beyer H-G, Standish R, Kendall G, Wilson S, Harman M, Wegner J, Dasgupta D, Potter M, Schultz A, Dowsland K, Jonoska N, Miller J (eds) Genetic and Evolutionary Computation – GECCO 2003, Springer Verlag Berlin Heidelberg New York, Part I, pp. 207–218

Kerr S, Personaaz L, Dreyfus G (1992) Handwritten Digit Recognition by Neural Networks with Single Layer Training. IEEE Transactions on Neural Networks 3(5) November

Khosla R, Dillon T (1997) Engineering Intelligent Hybrid Multi-Agent Systems. Kluwer Academic Publishers, Boston Dodrecht London

Kiendl H (1994) The Inference Filter. In: Proceedings of 2nd European Congress on Intelligent Techniques and Soft Computing EUFIT'94, pp. 443–447

Korbicz J, Patan K, Obuchowitcz O (1999) Dynamic neural networks for process modeling in fault detection and isolation systems. J Applied Mathematics and Computer Science 3 (9): 519–546

Kosba E, Dimitrova V, Boyle R (2003) Using Fuzzy Techniques to Model Students in Web-Based Learning Environments. In: Palade V, Howlett J R, Jain L (eds) Knowledge-Based Intelligent Information and Engineering Systems. Springer-Verlag, Berlin Heidelberg New York, Part II, pp. 222–228

Koscielny J M, Syfert M, Bartys M (1999) Fuzzy-logic fault diagnosis of Industrial process actuators. J Applied Mathematics and Computer Science 3 (9): 637–652

Kosko B, Neural Networks and Fuzzy Systems, Prentice-Hall, Englewood Cliffs: NJ, 1992

Langari R, "Synthesis of nonlinear control strategies via fuzzy logic", in Proc. of American Control Conference, pp. 1855–1859, 1993

Langari R, Wang L (1996) Fuzzy models, modular networks, and hybrid learning, Fuzzy Sets and Systems 79, 141–150

Lee C, Hwang C, Shih P Y, (1994) A Combined Approach to Fuzzy Model Identification, in *IEEE Transaction on System, Man and Cybernetics*, vol. SMC-24, pp. 736–744

Lee K-M, Yamakawa T, Uchino E, Lee, Ke M, (1997) A Genetic Algorithm Approach to Job Shop Scheduling. In: Proceedings of ICONPI/ANZIIS/ANNES'97. Dunedin, Otago, New Zealand, pp. 1030–1033

Lee M A, Esbensen H (1997) Fuzzy/Multiobjective Genetic Systems for Intelligent Design Tools and Components. In: Pedrycz W (ed.), Fuzzy Evolutionary Computation. Kluwer Academic Publishers, pp. 57–81

Lee M A, Esbensen H, Lemaitre L (1995) The Design of Hybrid Fuzzy/Evolutionary Multiobjective Optimization Algorithms. In: Furuhashi T, Uchikawa Y (eds), Fuzzy Logic, Neural Networks and Evolutionary Computation. IEEE/Nagoya-University World Wiseperson Workshop, Nagoya, Japan, Springer Verlag, Berlin, Heidelberg, Germany, pp. 188–215

Lee M A, Takagi H (1993) Integrating Design Stages of Fuzzy Systems Using Genetic Algorithms. In: Proc IEEE Int Conf on Fuzzy Systems (FUZZ-IEEEE '93), San Francisco California, pp. 612–617

Lee M A, Takagi H (1994) Learning Control Strategies for High Performance Genetic Algorithms. In: Proceedings of IEEE/Nagoya University WWW Conference on Fuzzy Logic and Neural Networks/Genetic Algorithms, August, Nagoya, Japan, pp. 93–96

Lee S S, "Predicting atmospheric ozone using neural networks as compared to some statistical methods", in Proc. of the IEEE Technical Applications Conference and Workshops, NORTHCON'95, pp. 309–316, 1995

Lekkas G P, Avouris N M, Viras L G, "Case-Based Reasoning in Environmental Monitoring Applications", Applied Artificial Intelligence, vol. 8(3), pp. 359–376, 1994

Lin C-T, Lee C S G, "Neural-Network-based Fuzzy Logic Control and Decision System", IEEE Trans. On Computers, vol. 40(12), pp. 1320–1336, 1991

Lin D S, Leou J-J (1997) A Genetic Algorithm Approach to Chinese Handwriting Normalization. J IEEE Transactions on Systems, Man and Cybernetics 27(6), pp. 999–1006

Mamdani E H, "Application of fuzzy logic to approximate reasoning using linguistic synthesis", IEEE Trans. Computers, vol. C-26(12), pp. 1192–1191, 1977

Marchal P, Nussbaum P, Piguet C, Durand S, Mange D, Sanchez E, Stauffer A, Tempesti G (1996) Embryonics: The Births of Synthetic Life. In:

Sanchez E. Tomassini M (eds) Towards Evolvable Hardware. Springer Verlag, Heidelberg, pp. 166–195

Marcu T, Mirea L, Frank P M (1999) Development of dynamic neural networks with application to observer-based fault detection and isolation. J of Applied Mathematics and Computer Science 3 (9): 547–570

Mayo M, Mitrovic A (2001) Optimising ITS Behaviour with Bayesian Networks and Decision Theory. J International of Artificial Intelligence in Education 12: 124–153

Mayo M, Mitrovic A, McKenzie J (2000) CAPIT: An Intelligent Tutoring System for Capitalisation and Punctuation. In: Kinshuk J C, Okamoto T (eds) Advanced Learning Technology: Design and Development Issues. IEEE Computer Society, Los Alamitos, CA, pp. 151–154

Mazzatorta P, Benfenati E, Neagu C-D, Gini G, The importance of scaling in data mining for toxicity prediction, Journal of Chemical Information and Computer Sciences, 2002; 42/5, 1250–1255, American Chemical Society, George W A Milne & Associate Editors

Mazzatorta P, Benfenati E, Neagu C-D, Gini G, Tuning neural and fuzzy-neural networks for toxicity modelling, Journal of Chemical Information and Computer Sciences, 2003; 43/2, 513–518, American Chemical Society, George W A Milne & Associate Editors

Mc Taggart J (2001) Intelligent Tutoring Systems and Education for the Future. In: 512X Literature Review April 30, pp. 2. http://www.drake.edu/mathcs/mctaggart/C1512X/LitReview.pdf

McCullagh J, Choi B, Bluff K (1997) Genetic Evolution of a Neural Network's Input Vector for Meteorological Estimation. In: Kasabov N, Kozma R, Ko K, Coghill G, Gedeon T (eds) Progress in Connectionist-Based Information Systems, Springer Verlag, pp. 1046–1049

Mihaila D, Fagarasan F, Negoita M Gh (1996) Architectural Implications of Genetic Algorithms Complexity in Evolvable Hardware Implementation. In: Proceedings of the European Congress EUFIT'96. Vol. 1. September, Aachen, Germany, pp. 400–404

Mills A P Jr, Turberfield M, Turberfield A J, Yurke B, Platzman P M (2001) Experimental Aspects of DNA Neural Network Computation., In: J Soft Computing 5: 10–18, Springer Verlag

Murray T (1999) Authoring Intelligent Tutoring Systems: An Analysis of the State of the Art. J International Journal of Artificial Intelligence 10: 98–129

Murthy S K, Kasif S, Salzberg S, "A System for Induction of Oblique Decision Trees", Journal of Artificial Intelligence Research, vol. 2, pp. 1–32, 1994

Nasraoui O, Gonzales F, Cardona C, Rojas C, Dasgupta D (2003) A Scalable Artificial Immune System Model for Dynamic Unsupervised Learning. In: Cantu-Paz E, Foster J, Kalaynmoy D, Lawrence D D, Rajkumar R, O'Reilly U-M, Beyer H-G, Standish R, Kendall G, Wilson S, Harman M, Wegner J, Dasgupta D, Potter M, Schultz A, Dowsland K, Jonoska

N, Miller J (eds) Genetic and Evolutionary Computation – GECCO 2003, Springer Verlag Berlin Heidelberg New York, Part I, pp. 219–230

Nauck D, Kruse R (1998) Nefclass – a soft computing tool to build readable fuzzy classifiers. J BT Technol 16 (3): 89–103

Nauck D, Kruse R, 1998. NEFCLASS-X: A Neuro-Fuzzy Tool to Build Readable Fuzzy Classifiers. *BT Tech. J.* 16(3):180–192

Neagu C-D, 2000, Modular neuro-fuzzy networks: solutions for explicit and implicit knowledge integration, in The Annals of University of Galati, Fascicle III, 52–59

Neagu C-D, 2002, Toxicity prediction using assemblies of hybrid fuzzy neural models, Proceedings of the 6th International Conference on Knowledge-Based Intelligent Information & Engineering Systems – KES2002, Milan – Italy, 16–18 Sept. 2002, 1093–1098, IOS Press, ISBN 1-58603-280-1

Neagu C-D, AO Aptula, G. Gini, Neural and Neuro-Fuzzy Models of Toxic Action of Phenols, IEEE Int'l Symp Intelligent Systems: Methodology, Models, Applications in Emerging Technologies IS2002, Varna, Bulgaria, 10–12 Sept. 2002, 283–288

Neagu C-D, Benfenati E, Gini G, Mazzatorta P, Roncaglioni A, Neuro-Fuzzy Knowledge Representation for Toxicity Prediction of Organic Compounds, 15th European Conference on Artificial Intelligence ECAI2002, July 21–26, 2002, Lyon, France, 498–502, IOS Press

Neagu C-D, Bumbaru S, "Explicit Knowledge Representation using Multi Purpose Neural Networks", in Proc. of the International Conference on Control Systems and Computer Science CSCS12, edited by I. Dumitrache, M. Dobre, Bucharest, vol. 2, pp. 37–42, 1999

Neagu C-D, Gini G, 2003, "Neuro-Fuzzy Knowledge Integration applied in Toxicity Prediction", chapter 12 in *Innovations in Knowledge Engineering* (Jain R, Abraham A, Faucher C, van der Zwaag B J: eds), Advanced Knowledge International

Neagu C-D, Gini G, "Neuro-Fuzzy Knowledge Integration applied in Toxicity Prediction", chapter 12 in Innovations In Knowledge Engineering (Ravi Jain, Ajith Abraham, Colette Faucher and Berend Jan van der Zwaag: eds), 311–342, ISBN: 0-9751004-0-8, Advanced Knowledge International, 2003

Neagu C-D, Negoita M, Palade V, "Aspects of Integration of Explicit and Implicit Knowledge in Connectionist Expert Systems", in Proc. of ICONIP'99, Perth, vol. 2, pp. 759–764, 1999

Neagu C-D, Avouris N M, Kalapanidas E, Palade V, Neurosymbolic Integration in a Knowledge-based System for Air Quality Prediction, Applied Intelligence, 2002; 17/2: 141–169, Kluwer

Neagu C-D, Bumbaru S, An Interactive Fuzzy Operator Used To Interpret Connectionist Knowledge, Journal of Control Engineering and Applied Informatics, 2001; 3/3: 53–59

Neagu C-D, Palade V, Fuzzy Computing in a Multi Purpose Neural Network Implementation, LNCS 1625 -- Computational Intelligence: Theory and Applications (Bernd Reusch: ed.), 1999; 697–700, Springer Verlag

Neagu C-D, Palade V, 2003, A Neuro-Fuzzy Approach for Functional Genomics Data Interpretation and Analysis, Journal of Neural Computing and Applications, Springer-Verlag **12/3–4**, 153–159

Neagu C-D, Palade V, An interactive fuzzy operator used in rule extraction from neural networks, Neural Network World: (Novak M, Ed.), 2000; 10/4, 675–684

Neagu C-D, Palade V, Modular neuro-fuzzy networks: an overview of explicit and implicit knowledge integration, the 15th Int'l FLAIRS-02 Conference, Special Track on Integrated Intelligent Systems, Pensacola, Florida 14–16 May 2002, 277–281, AAAI Press

Neagu C-D, Palade V, "A Neuro-Fuzzy Approach To Photochemical Pollution Prediction", in Proc. of International Symposium On Systems Theory, Automation, Robotics, Computers, Informatics, Electronics And Instrumentation, SINTES10, Craiova, pp. 128–130, 2000

Neagu C-D, Palade V, "Fuzzy Computing in a Multi Purpose Neural Network Implementation", in Proc. of the International Conference 6th Fuzzy Days in Dortmund, edited by Reusch B, Springer Verlag, pp. 697–700, 1999

Neagu C-D, *Hybrid Intelligent Systems: Combination Systems*, Bucharest: Matrix Rom, 2001

Neagu CD, Palade V (2003) Hybrid Intelligent Systems (in Romanian). Matrix Rom, Romania

Negoita M Gh (1997) A Modern Solution for the Technology of Preventing and Alarm Systems: Evolutionary Computation in Evolvable Hardware Implementation. In: Proceedings of The Second Dunav Preving International Conference on Preventing and Alarm Systems. Belgrade, November, pp. 201–209

Negoita M Gh (1997) Evolutionary Computation in Evolvable Hardware Implementation, An Advanced Topic Lecture. The University of Auckland, Auckland, New Zealand, December

Negoita M Gh, Dediu A-H, Mihaila D (1997) Application of GA with Local Improvement of the Chromosomes for the Design of EHW Architectures. In: Proceedings of EUFIT'97. Aachen, Germany, pp. 814–818

Negoita M Gh (1994) Genetic Algorithms in Soft-Computing Environment. Theory and Applications. A Tutorial Presentation, Gerhardt Mercator University, Duisburg, Germany, October

Negoita M Gh (1995) Evolutionary Computation in the Soft Computing Framework. In: Zimmermann H-Z, Negoita M Gh, Dascalu D (eds) Real World Applications of Intelligent Technologies. Editura Academiei Romane, Bucharest, Romania, pp. 113–139

Negoita M Gh (1996) Methods Based on Evolutionary Computation Techniques for Implementation of Evolvable Hardware. In: Proceedings of the

International Conference of Technical Informatics – vol. 1– Computer Science and Engineering. Timisoara, Romania, pp. 37–45

Negoita M Gh (2002) Intelligent Multi-Agent Hybrid Systems (IMAHS). A Course for Internal Use of Students Only, Wellington Institute of Technology, Wellington, New Zealand

Negoita M Gh (2002a) Evolutionary Computation in Evolvable Hardware Implementation – Implication on Engineering Design and Automation. A Tutorial Presentation at AI'02, Canberra, Australia, December

Negoita M Gh, Agapie A, Fagarasan F (1994b) The Fusion of Genetic Algorithms and Fuzzy Logic. Application in Expert Systems and Intelligent Control. In: Proceedings of IEEE/Nagoya University WWW Conference on Fuzzy Logic and Neural Networks/Genetic Algorithms, August, Nagoya, Japan, pp. 130–133

Negoita M Gh, Arotaritei D (2003) A GA with Variable-Length Chromosomes for Optimization Objectives of Fuzzy Recurrent Neural Networks. In: Barry A (ed) 2003 Genetic and Evolutionary Computation Conference Workshop Program, University of Bath, pp. 208–214

Negoita M Gh, Fagarasan F, Agapie A (1994a) Soft-Computing: Fusion Examples of Intelligent Techniques. In: Proceedings of International Conference OPTIM'94, Transilvania University of Brasov, Romania, May, pp. 335–340

Negoita M Gh, Fagarasan F, Agapie A (1994c) Application of Genetic Algorithms in Solving Fuzzy Relational Equations. In: Proceedings of EU-FIT'94, Aachen, Germany, September, pp. 1126–1129

Negoita M Gh, Giuclea M, Dediu A-H (1995) GA to Optimize Approximate Solutions of Fuzzy Relational Equations for Fuzzy Systems or Controllers. In: Proceedings of ANNES'95, Dunedin, Otago, New Zealand, pp. 124–127

Negoita M Gh, Mihaila D (1995) Intelligent Techniques Based on Genetic Evolution with Applications to Neural Networks Weights Optimization. In: Proceedings of 14-th International Congress on Cybernetics. Namur, Belgium

Negoita M Gh, Pritchard D (2003a) Testing Intelligent Tutoring Systems by Virtual Students. In: Arif Wani M (ed) Proc Int Conf on Machine-Learning and Applications (ICMLA'03), Los Angeles, USA, pp. 98–104

Negoita M Gh, Pritchard D (2003b) Some Test Problems Regarding Intelligent Tutoring Systems. In: Palade V, Howlett J R, Jain L (eds) Knowledge-Based Intelligent Information and Engineering Systems. Springer-Verlag, Berlin Heidelberg New York, Part II, pp. 98–992

Negoita M Gh, Pritchard D (2004a) Using Virtual Student Model for Testing Intelligent Tutoring Systems. J Interactive Technology & Smart Education 1: 3–10

Negoita M Gh, Pritchard D (2004b) A "Virtual Student" Leads to the Possibility of Optimizer Agents in an ITS. In: Kantardzic M (ed) Proc. ICMLA'04, Louisville, KY, USA, in press

Negoita M Gh, Stoica A (2004) EHW in the Framework of Computational Intelligence – Implications on Engineering Design and Intelligence of Autonomous Systems A Tutorial Lecture KBCS-04 International Conference on Artificial Intelligence, Hyderabad, India, December, 17–22

O' Riordan C, Griffith J (2003) Providing Personalised Recommendations in a Web-Based Education System. In: Palade V, Howlett J R, Jain L (eds) Knowledge-Based Intelligent Information and Engineering Systems. Springer-Verlag, Berlin Heidelberg New York, Part II, pp. 245–251

Oda T, White T (2003) Developing an Immunity to Spam. In: Cantu-Paz E, Foster J, Kalaynmoy D, Lawrence D D, Rajkumar R, O'Reilly U-M, Beyer H-G, Standish R, Kendall G, Wilson S, Harman M, Wegner J, Dasgupta D, Potter M, Schultz A, Dowsland K, Jonoska N, Miller J (eds) Genetic and Evolutionary Computation – GECCO 2003, Springer Verlag Berlin Heidelberg New York, Part I, pp. 207–242

Oeda S, Ichimura T, Yamashita T, Yoshida K (2003) A Proposal of Immune Multi-agent Neural Networks and Its Application to Medical Diagnostic System for Hepatobiliary Disorders. In: Palade V, Howlett J R, Jain L (eds) Knowledge-Based Intelligent Information and Engineering Systems. Springer-Verlag, Berlin Heidelberg New York, Part II, pp. 526–532

Ohlsson S (1987) Some Principles of Intelligent Tutoring. In: Lawler, Yazdani (eds) Artificial Intelligence and Education. Ablex Norwood N J, vol 1, pp. 203–238

Okamoto T, Watanabe T, Ishida Y (2003) Towards an Immunity-Based System for Detecting Masqueraders. In: Palade V, Howlett J R, Jain L (eds) Knowledge-Based Intelligent Information and Engineering Systems. Springer-Verlag, Berlin Heidelberg New York, Part II, pp. 488–495

Omlin C W, Giles C L, "Extraction of Rules from Discrete-Time Recurrent Neural Networks", Neural Networks, vol. 9, pp. 41–52, 1996

Orey M A, Nelson W A (1993) Development Principles for Intelligent Tutoring Systems: Integrating Cognitive Theory into the Development of Computer-based Instruction. J Educational Technology Research and Development 41(1): 59–72

Ortega C, Mange D, Smith S, Tyrrell A (2000) Embryonics: A Bio-Inspired Cellular Architecture with Fault-Tolerant Properties. In: J Genetic Programming and Evolvable Machines 1–3: 187–215

Palade V, Bumabaru S, Negoita M (1998) A Method for Compiling Neural Networks into Fuzzy Rules Using Genetic Algorithms and Hierarchical Approach. In: Proc KES Int Conf on Knowledge-Based Intelligent Electronic Systems, Adelaide, Australia, pp. 353–358

Palade V, Negoita M Gh, Ariton V (1999) Genetic Algorithms Optimization of Knowledge Extraction from Neural Networks. In: Proceedings of ICONIP'99, November, Perth, Australia, pp. 752–758

Palade V, Patton R J, Uppal F J, Quevedo J, Daley S (2002a) Fault diagnosis of an industrial gas turbine using neuro-fuzzy methods. In: Proceedings of the 15th IFAC World Congress. Barcelona-Spain, 21-26 July, pp. 2477–2482

Palade V, "GA Optimization of Knowledge Extraction from Neural Networks", in Proc. of ICONIP99, Sidney, 1999

Park D, Kandel A, Langholz G (1994) Genetic Based New Fuzzy Reasoning Method with Application to Fuzzy Control. IEEE Trans. on SMC 1: 39–47

Patton R J, Simani S (1999) Identification and fault diagnosis of a simulated model of an industrial gas turbine. Technical Research Report. The University of Hull-UK, Department of Engineering

Patton R J, Lopez-Toribio C J, Uppal F J (1999) Artificial Intelligence Approaches to fault diagnosis for dynamic systems. J Applied Mathematics and Computer Science 3 (9): 471–518

Patton R J, Simani S, Daley S and Pike A (2000). Fault diagnosis of a simulated model of an industrial gas turbine prototype using identification techniques. In: Proceedings of the 4th IFAC Symposium on Fault Detection, Supervision and Safety for Technical Processes, pp. 518–523, Budapest.

Paun Gh, Rozenberg G, Salomaa A (1998) DNA Computing – New Computing Paradigms. Springer-Verlag, Berlin Heidelberg New York

Pedrycz W (1990) Algorithms for Solving Fuzzy Relational Equations in a Probabilistic Setting. In: Fuzzy Sets and Systems 38

Pedrycz W, da Rocha A F, "Fuzzy-Set Based Models of Neurons and Knowledge-Based Networks", IEEE Transactions on Fuzzy Systems, vol.I(5), pp. 254–266, 1993

Pedrycz W, "Fuzzy Neural Networks and Neurocomputations", Fuzzy Sets and Systems, vol. 56, pp. 1–28, 1993

Prem E, Mackinger M, Dorffner G, Porenta G, Sochor H, "Concept Support as a Method for Programming Neural Networks with Symbolic Knowledge", TR-93-04, OEFAI, 1993

Pritchard D, Negoita Gh M (2004) A Fuzzy – GA Hybrid Technique for Optimisation of Teaching Sequences Presented in ITSs. In: Reusch B (ed) Proc 8-th Fuzzy Days. LNCS, Springer Verlag, Berlin Heidelberg New York, in press

Ranawana R, Palade V (2004) A Neural Network Based Multi-Classifier System for Gene Identification in DNA Sequences. J Neural Computing, in press

Redfield L C, Steuck K (1991) The Future of Intelligent Tutoring Systems. In: Burns, Hugh, Parlett, James W, Redfield L C (eds) Intelligent Tutoring Systems: Evolutions in Design, pp. 265–284

Rose A J, Deaton R, Franceschetti D R, Garzon M, Stevens E S Jr (1999) A Statistical Mechanical Treatment of Error in the Annealing Biostep of DNA Computation. Special Program in DNA and Molecular Computing at GECCO – 99. In Proceedings. Morgan – Kaufman, Orlando Florida, pp. 1829–1834

Rumelhart D E, and McClelland J L, *Parallel Distributed Processing, Explanations in the Microstructure of Cognition*, MIT Press, 1986

Sanchez E (1970) Resolution of composite relational equations. In: Information and Control 30

Schultz T W, *Tetrahymena* in aquatic toxicology: QSARs and ecological hazard assessment, in: Pauli W, Berger S (eds.), Procs of the Int.l Workshop on a Protozoan Test Protocol with Tetrahymena in Aquatic Toxicity Testing. Federal Environmental Agency tech.rep. no. 106 99 999/03, Berlin 1996, 31–66

Schultz T W, Cronin M T D, Walker J D, Aptula A O, 2003, Quantitative structure-activity relationships (QSARs) in toxicology: a historical perspective. Journal of Molecular Structure (Theochem) 622: 1–22

Shafter G (1986) Probability Judgment in Artificial Intelligence. In: Kanal L, Lemmer J (eds) Uncertainty in Artificial Intelligence. North-Holland, New York

Shann J J, Fu H C (1995) A fuzzy neural network for rule acquiring on fuzzy control systems. J Fuzzy Sets and Systems 71: 345–357

Shimojima K, Fukuda T, Hasegawa I (1995) Self-tuning Fuzzy Modeling with Adaptive Membership Function, Rules, and Hierarchical Structure Based on Genetic Algorithm. J Fuzzy Sets and Systems 71: 294–309

Shimonara K (1994) Evolutionary Systems for Brain Communications – Towards an Artificial Intelligence. In: Brooks A R, Maes P (eds) Artificial Life IV. The MIT Press, pp. 4–7

Shimooka T, Shimizu K (2003) Idiotypic Network Model for Feature Extraction in Pattern Recognition. In: Palade V, Howlett J R, Jain L (eds) Knowledge-Based Intelligent Information and Engineering Systems. Springer-Verlag, Berlin Heidelberg New York, PartI I, pp. 511–518

Sima J, Cervenka J, "Neural Knowledge Processing in Expert Systems", TR no. V-735, Institute of Computer Science, Academy of Sciences of the Czech Republic, 1997

Smyth B, (2003) Intelligent Navigation on the Mobile Internet. In: Palade V, Howlett J R, Jain L (eds) Knowledge-Based Intelligent Information and Engineering Systems. Springer-Verlag, Berlin Heidelberg New York, Part I, pp. 17–19

Sucar L E, Perez-Brito J, Ruiz-Suarez J C, Morales E, "Learning Structure from Data and Its Application to Ozone Prediction", Applied Intelligence, vol. 7, pp. 327–338, 1997

Sun R (1994) CONSYDERR: A Two Level Hybrid Architecture Structuring Knowledge for Commomsense Reasoning, in Proc. of the First International Symposium on Integrating Knowledge and Neural Heuristics, Florida, USA, 32–39

Sugeno M, Kang G T, "Structure identification of fuzzy model", Fuzzy Sets and Systems, vol. 28, pp. 15–33, 1988

Tachibana K, Furuhashi T (1994) A hierarchical fuzzy modelling method using genetic algorithm for identification of concise submodels. In: Proc of 2nd Int Conference on Knowledge-Based Intelligent Electronic Systems. April, Adelaide, Australia

Takagi H (1996) Industrial and Commercial Applications of NN/FS/GA/ Chaos in 1990s. In: Proceedings of International Workshop on Soft Computing in Industry (IWSCI'96). April, Muroran, Hokkaido, Japan, pp. 160–165

Takagi H, "Cooperative system of neural networks and fuzzy logic and its application to consumer products", in Industrial Applications of Fuzzy Control and Intelligent Systems, edited by Yen J, Langari R, Zadeh L A, Van Nostrand Reinhold: NY, 1994

Takagi T, Sugeno M, Fuzzy identification of systems and its applications to modeling and control, IEEE Trans. on System, Man, and Cybernetics, 1985; 15: 116–132

Takeda F, Omatu S, Onami S, Kadono T, Terada, K (1994) A Paper Currency Recognition Method by a Neural Network Using Masks and Mask Optimization by GA. In: Proceedings of IEEE/Nagoya University WWW Conference on Fuzzy Logic and Neural Networks/Genetic Algorithms, August, Nagoya, Japan, pp. 125–129

Teodorescu H-N (1994) Non-Linear Systems, Fuzzy Systems, and Neural Networks. In: Proceedings of 3rd Conference on Fuzzy Logic, Neural Networks and Soft Computing, pp. 17–28

Thompson A (1995) Evolving Fault Tolerant Systems CSRP 385. In: Proceedings of 1st IEE/IEEE IC on GA's in Eng Sys GALESIA'95, pp. 524–529

Thrun S B (1994) Extracting Symbolic Knowledge from Artificial Neural Networks. Revised version of *Technical Report TR-IAI-93-5*, Institut fuer Informatik III – Universitaet Bonn

Tong R M (1978) Synthesis of Fuzzy Models for Industrial Process. Int J Gen Syst 4: 143–162

Torresen J (1997) Evolvable Hardware – The Coming Hardware Design Method?, In: Kasabov N(Ed) Neuro-fuzzy Tools and Techniques, Springer Verlag

Uppal F J, Patton R J, Palade V (2002) Neuro-fuzzy based fault diagnosis applied to an electro-pneumatic valve. In: Proceedings of the 15th IFAC World Congress, Barcelona-Spain, 21-26 July, pp. 2483–2488

Vallejo E E, Ramos F (2003) Evolutionary Two-Dimensional DNA Sequence Alignment. In: Cantu-Paz E, Foster J, Kalaynmoy D, Lawrence D D, Rajkumar R, O'Reilly U-M, Beyer H-G, Standish R, Kendall G, Wilson S, Harman M, Wegner J, Dasgupta D, Potter M, Schultz A, Dowsland K, Jonoska N, Miller J (eds) Genetic and Evolutionary Computation – GECCO 2003, Springer Verlag Berlin Heidelberg New York, Part I, pp. 429–430

Wang D, Fang S-C, (1997) A Genetics-Based Approach for Aggregated Production Planning in A Fuzzy Environment in *IEEE Transactions on Systems, Man and Cybernetics* vol. 27, no. 5, September, pp. 636–645

Wasiewicz P, Mulawka J J (2001) Molecular Genetic Programming. In: J Soft Computing 5: 106–113, Springer Verlag

Watanabe Y, Ishida Y (2003) Immunity-Based Approaches for Self-Monitoring in Distributed Intrusion Detection System. In: Palade V, Howlett J R, Jain L (eds) Knowledge-Based Intelligent Information and Engineering Systems. Springer-Verlag, Berlin Heidelberg New York, Part II, pp. 503–510

Wenger E (1987) Artificial Intelligence and Tutoring Systems. Morgan Kaufmann, Los Altos CA

Wermter S, Sun R, Hybrid Neural Systems, Springer Verlag: Heidelberg, 2000

West M, Garzon H M, Blain D (2003) DNA-Like Genomes for Evolution *in silico*. In: Cantu-Paz E, Foster J, Kalaynmoy D, Lawrence D D, Rajkumar R, O'Reilly U-M, Beyer H-G, Standish R, Kendall G, Wilson S, Harman M, Wegner J, Dasgupta D, Potter M, Schultz A, Dowsland K, Jonoska N, Miller J (eds) Genetic and Evolutionary Computation – GECCO 2003, Springer Verlag Berlin Heidelberg New York, Part I, pp. 412–423

WHO (World Health Organization), Air quality guidelines for Europe, WHO: Copenhagen, 1987

Williams R J, Zipser D (1989) Experimental Analysis of the Real-Time Recurrent Learning Algorithm. J Connection Science 1: 87–111

Yager R R, Filev D P, Sadeghi T (1994) Analysis of Flexible Structured Fuzzy Logic Controllers. IEEE Trans on SMC 7:1035–1043

Yang X, Furuhashi T, Obata K, Uchikawa Y (1995) Constructing a High Performance Signature Verification System Using A GA Method. In: Proceedings of ANNES'95, Dunedin, Otago, New Zealand, pp. 170–173

Yoshida K, Matsumoto K, Nakada K, Akutsu T, Fujii S, Ichimura H (2003) A Trial of a Bidirectional Learning Management Tool for Promoting Learning by Mobile Phone. In: Palade V, Howlett J R, Jain L (eds) Knowledge-Based Intelligent Information and Engineering Systems. Springer-Verlag, Berlin Heidelberg New York, Part II, pp. 757–763

Yoshino T, Yoshinori F, Munemori J (2003) Data Processing Method of Small-Scale Five Senses Communication System. In: Palade V, Howlett J R, Jain L (eds) Knowledge-Based Intelligent Information and Engineering Systems. Springer-Verlag, Berlin Heidelberg New York, Part II, pp. 748–753

Zadeh L A (1983) The Role of Fuzzy Logic in the Management of Uncertainty in Expert Systems. J Fuzzy Sets and Systems 11: 199–227

Zadeh L A (2003) From Search Engines to Question-Answering Systems – The Need For New Tools. URL: http://www-bisc.cs.berkeley.edu Computer Science Division Department of EECS UC Berkeley

Zadeh L A, Nikravesh M (2002) Perception-Based Intelligent Decision Systems. ONR Summer 2002 Program Review, URL: http://www-bisc.cs.berkeley.edu Computer Science Division Department of EECS UC Berkeley

Zadeh L A, "The role of fuzzy logic in the management of uncertainty in expert systems", Fuzzy Sets and Systems, vol. 11(3), pp. 199–227, 1983

Zhang J and Morris J (1996). Process modeling and fault diagnosis using fuzzy neural networks, *Fuzzy Sets and Systems* vol. 79, pp. 127–140

Zickus M, "Influence Of Meteorological Parameters On The Urban Air Pollution And Its Forecast", PhD Thesis, University of Vilnius, 1999

Zimmermann H-J (1991) Fuzzy Set Theory and Its Applications (Second Revised Edition). Kluwer Academic Publishers, Boston, Dodrecht, London

Zimmermann H-J, Negoita Gh M, Dascalu D (Eds) (1996) Real World Application of Intelligent Technologies. Editura Academiei Romane, Bucharest